PRINCIPLES OF THERMAL
ANALYSIS AND CALORIMETRY

RSC Paperbacks

RSC Paperbacks are a series of inexpensive texts suitable for teachers and students and give a clear, readable introduction to selected topics in chemistry. They should also appeal to the general chemist. For further information on all available titles contact:

Sales and Customer Care Department, Royal Society of Chemistry,
Thomas Graham House, Science Park, Milton Road, Cambridge CB4 0WF, UK
Telephone: +44 (0)1223 432360; Fax: +44 (0)1223 423429; E-mail: sales+rsc.org

Recent Titles Available

The Chemistry of Fragrances
compiled by David Pybus and Charles Sell
Polymers and the Environment
by Gerald Scott
Brewing
by Ian S. Hornsey
The Chemistry of Fireworks
by Michael S. Russell
Water (Second Edition): A Matrix of Life
by Felix Franks
The Science of Chocolate
by Stephen T. Beckett
The Science of Sugar Confectionery
by W. P. Edwards
Colour Chemistry
by R. M. Christie
Understanding Batteries
by Ronald M. Dell and David A. J. Rand
Principles of Thermal Analysis and Calorimetry
Edited by P. J. Haines

Future titles may be obtained immediately on publication by placing a standing order for RSC Paperbacks. Information on this is available from the address above.

RSC Paperbacks

PRINCIPLES OF THERMAL ANALYSIS AND CALORIMETRY

Edited by P. J. Haines

*Oakland Analytical Services, Farnham,
Surrey, UK*

Contributors

G. R. Heal
University of Salford, UK

P. G. Laye
University of Huddersfield, UK

D. M. Price
Loughborough University, UK

S. B. Warrington
Loughborough University, UK

R. J. Wilson
GlaxoSmithKline, Harlow, UK

ROYAL SOCIETY OF CHEMISTRY

ISBN 0-85404-610-0

A catalogue record for this book is available from the British Library

Published by The Royal Society of Chemistry,
Thomas Graham House, Science Park, Milton Road,
Cambridge CB4 0WF, UK
Registered Charity Number 207890

For further information see our web site at www.rsc.org

Typeset in Great Britain by Vision Typesetting, Manchester, UK
Printed in Great Britain by TJ International Ltd, Padstow, Cornwall

Foreword

The Thermal Methods Group of the Royal Society of Chemistry, which was founded in 1965, has a tradition of education in thermal analysis dating back to its first residential thermal analysis school held at the Cement and Concrete Research Association in 1968. The Group has continued to be at the forefront of thermal education through the organisation of schools, specialist meetings and both national and international conferences.

Over the past twenty years, thermal methods have seen a rapid growth in their use in an increasingly wide range of applications. In addition, a number of powerful new techniques have been developed recently. It is therefore timely that a group of UK scientists have pooled their specialist expertise to produce this wide-ranging book, which should be of considerable value to those who are new to the field or who are coming to a particular technique for the first time. The broad range of techniques and applications covered means that there is also much to interest the more experienced thermal analyst.

Throughout most of its long life the Thermal Methods Group has been fortunate in having an outstanding contribution from three of its members, namely Professor David Dollimore (Chairman 1969–1971), Dr Cyril J. Keattch (Hon. Secretary 1965–1998) and Dr Robert C. Mackenzie (Chairman 1965–1967). These scientists throughout their long and distinguished careers were unstinting in helping young workers and those new to the field to develop their thermal analysis expertise. It is a most fitting tribute that this book is dedicated to their memory and to their invaluable contribution to the development of thermal analysis.

Edward L. Charsley
Past President of the International Confederation
for Thermal Analysis and Calorimetry, (ICTAC)
Centre for Thermal Studies, University of Huddersfield, UK

Dedicated to the memory of
Dr Cyril Jack Keattch,
1928–1999
Honorary Secretary of the Thermal Methods Group
for its first 33 years

Dr Robert Cameron Mackenzie
1920–2000
Founder Member of the TMG (Chairman 1965–1967) and ICTAC

Professor David Dollimore
1927–2000
Chairman of the TMG 1969–1971

C. J. Keattch   R. C. Mackenzie   D. Dollimore

Contents

Chapter 4
Thermomechanical, Dynamic Mechanical and Dielectric Methods   94
D. M. Price

Acknowledgements

The authors of this text acknowledge their debt to Cyril Keattch, Robert Mackenzie and David Dollimore.

It is also a pleasure to acknowledge the contribution made by the instrument manufacturers over many years. They are listed in the Appendices, and the brochures, application notes and personal help which they have given, and continue to give, plays a vital part in the use of thermal and calorimetric analysis. The Group gratefully acknowledges their permission to use several of the diagrams in this text.

The Thermal Methods Group maintains a web site, through the Royal Society of Chemistry at **http://thermalmethodsgroup.org.uk** and a list server whereby requests for information and queries about techniques may be exchanged. If you wish to join the TMG Internet Newsgroup, please follow the intructions on the TMG web site.

Thanks are due to many people, particularly Dr Trevor Lever and Dr Michael Richardson for their help in preparing this book.

Peter J. Haines (*Editor*)

Contributors

P. J. Haines, *Oakland Analytical Services, 38 Oakland Avenue, Farnham, Surrey GU9 9DX, UK*

G. R. Heal, *Department of Chemistry and Applied Chemistry, University of Salford, Salford M5 4WT, UK*

P. G. Laye, *Centre for Thermal Studies, University of Huddersfield, Queensgate, Huddersfield HD1 3DH, UK*

D. M. Price, *Institute of Polymer Technology and Materials Engineering, Loughborough University, Loughborough LE11 3TU, UK*

S. B. Warrington, *Institute of Polymer Technology and Materials Engineering, Loughborough University, Loughborough LE11 3TU, UK*

R. J. Willson, *GlaxoSmithKline, New Frontiers Science Park (South), Harlow, Essex CM19 5AW, UK*

Chapter 1

Introduction

P. J. Haines

Oakland Analytical Services, Farnham, UK

MATERIALS, HEAT AND CHANGES

Whenever a sample of material is to be studied, one of the easiest tests to perform is to heat it. The observation of the behaviour of the sample and the quantitative measurement of the changes on heating can yield a great deal of useful information on the nature of the material.

In the simplest case, the temperature of the sample may increase, without any change of form or chemical reaction taking place. In short, it gets hotter. For many other materials, the behaviour is more complex. When ice is heated, it melts at $0\,°C$ and then boils at $100\,°C$. When sugar is heated, it melts, and then forms brown caramel. Heating coal produces inflammable gases, tars and coke. The list is endless, since every material behaves in a characteristic way when heated.

Thermal methods of analysis have developed out of the scientific study of the changes in the properties of a sample which occur on heating. Calorimetric methods measure heat changes.

Some sample properties may be obvious to the analyst, such as colour, shape and dimensions or may be measured easily, such as mass, density and mechanical strength. There are also properties which depend on the bonding, molecular structure and nature of the material. These include the thermodynamic properties such as heat capacity, enthalpy and entropy and also the structural and molecular properties which determine the X-ray diffraction and spectrometric behaviour.

Transformations which change the materials in a system will alter one or more of these properties. The change may be *physical* such as melting, crystalline transition or vaporisation or it may be *chemical* involving a

1

reaction which alters the chemical structure of the material. Even *biological* processes such as metabolism, interaction or decomposition may be included.

Sometimes a change brought about by heating may be reversed by cooling a sample afterwards. A pure organic substance melts sharply, for example benzoic acid melts at $122\,^{\circ}C$ and it recrystallises sharply when cooled below this temperature. Ammonium chloride dissociates into ammonia and hydrogen chloride gases when heated, but these recombine on cooling. At high temperature, calcium carbonate splits up to yield calcium oxide and carbon dioxide gas, and these too will recombine on cooling if the carbon dioxide is not removed. The system reaches an *equilibrium state* at a particular temperature.

$$CaCO_3 \text{ (solid)} \underset{\text{cool}}{\overset{\text{heat}}{\rightleftharpoons}} CaO \text{ (solid)} + CO_2 \text{ (gas)}$$

To raise the *temperature* of any system heat energy must be supplied and when sufficient energy is available the system will change into a more stable state. The mechanical properties of a material change as it is heated. Often it expands and becomes more pliable well below the melting point. These are fundamental, important changes on a molecular level, and their study enables the analyst to draw valuable conclusions about the sample, its previous history, its preparation, chemical nature and the likely behaviour during its proposed use.

The temperature at which a particular event occurs, or the temperature range over which a reaction happens, are often characteristic of the nature and history of a sample, and sometimes of the methods used to study it. Sharp transitions, such as the melting of pure materials, may be used to *calibrate* equipment and as the "fixed points" of thermometry and of the International Practical Temperature Scale (IPTS).

For example, how does the simple, pure inorganic compound potassium nitrate, KNO_3, behave when heated? At room temperature, say $20\,^{\circ}C$, this is a white, crystalline solid. To raise its temperature to $30\,^{\circ}C$ at constant presssure, we must supply an amount of heat depending on the specific heat capacity, C_p, approximately $1\,J\,K^{-1}\,g^{-1}$ at this temperature, the mass m of the sample and the change in temperature. So, for 1 g heated $10\,^{\circ}C$, we must supply $10\,J$. To complicate matters, the heat capacity changes with temperature as well. When the temperature reaches $128\,^{\circ}C$, the crystals change their structure, and this needs more energy, about $53\,J\,g^{-1}$. Then the new crystals are heated, when $C_p \approx 1.2\,J\,K^{-1}\,g^{-1}$, until the melting point of $338\,^{\circ}C$, when more heat must be supplied to melt the sample. Raising the temperature above the melting point eventually causes the sample to decompose to form potassium nitrite, KNO_2, so that the mass of the sample is decreased by around 16% and oxygen gas is given off.

This example illustrates the importance of thermal techniques and measurements. Calorimetry measures the amounts of heat, while appropriate thermal methods give the temperatures of phase changes, the temperatures of decomposition and the products of the reaction. Other methods will show the expansion, mass and colour changes on heating.

The analysis of thermal events may be approached in two ways, which overlap considerably. *Either* the experiment may be designed to measure thermal properties (heat capacity, enthalpy, entropy and free energy) with high precision and accuracy at particular temperatures and conditions, *or* we may study properties, including thermal properties, over a wider range of temperatures using a controlled heating procedure.

Which experiment is chosen depends on the sample to be analysed. There would be little point in obtaining highly accurate heat capacities on a polymeric or cement sample of complex composition, but its behaviour on heating would be informative. Theoretical work on organic structure and kinetics might require precise knowledge of equilibrium thermal properties which could not easily be obtained using variable temperature methods. Therefore, the techniques are complementary.

Since the worldwide adoption of the SI system of units it is perhaps useful to stress the symbols and units to be used for the physical quantities involved in these methods. The major quantities are given in Appendix 1A and the others may be found in the references.[1,2]

DEFINITIONS OF THERMAL AND CALORIMETRIC METHODS

Formal definitions are not essential, but those accepted by the scientific community may be found in the literature.[3,4]

Calorimetry is the measurement of the heat changes which occur during a process. The calorimetric experiment is conducted under particular, controlled conditions, for example, *either* at constant volume in a bomb calorimeter *or* at constant temperature in an isothermal calorimeter. Calorimetry encompasses a very large variety of techniques, including titration, flow, reaction and sorption, and is used to study reactions of all sorts of materials from pyrotechnics to pharmaceuticals.

Calorimetric methods may be classified *either* by the principle of measurement (*e.g.* heat compensating or heat accumulating), *or* by the method of operation (static, flow or scanning) *or* by the construction principle (single or twin cell). These will be discussed further in Chapter 5.

Thermal analysis is a group of techniques in which one (or more) property of a sample is studied while the sample is subjected to a controlled temperature *programme*. The *programme* may take many forms:

(a) The sample may be subjected to a constant heating (or cooling) rate ($dT/dt = \beta$), for example 10 K min^{-1}.
(b) The sample may be held isothermally ($\beta = 0$).
(c) A *"modulated temperature programme"* may be used where a sinusoidal or other alteration is superimposed onto the underlying heating rate.
(d) To simulate special industrial or other processes, a stepwise or complex programme may be used. For example, the sample might be equilibrated at 25°C for 10 min, heated at 10 K min^{-1} up to 200°C, held there for 30 min and then cooled at 5 K min^{-1} to 50°C.
(e) The heating may be controlled by the response of the sample itself.

THE FAMILY OF THERMAL METHODS

Every thermal method studies and measures a property as a function of temperature. The properties studied may include almost every physical or chemical property of the sample, or its products. The more frequently used thermal analysis techniques are shown in Table 1 together with the names most usually employed for them.

INSTRUMENTATION FOR THERMAL ANALYSIS AND CALORIMETRY

The modern instrumentation used for any experiment in thermal analysis or calorimetry is usually made up of four major parts:

- The sample and a container or holder;
- sensors to detect and measure a particular property of the sample and to measure temperature;
- an enclosure within which the experimental parameters (*e.g.* temperature, pressure, gas atmosphere) may be controlled;
- a computer to control the experimental parameters, such as the temperature programme, to collect the data from the sensors and to process the data to produce meaningful results and records.

This is shown schematically in Figure 1, and specific applications and instrumentation will be considered in the following chapters.

Table 1 *Thermal methods*

Technique	Abbreviation	Property	Uses
Thermogravimetry or (Thermogravimetric analysis)	TG TGA	Mass	Decompositions Oxidations
Differential thermal analysis	DTA	Temperature difference	Phase changes, reactions
Differential scanning calorimetry	DSC	Power difference or heat flow	Heat capacity, phase changes, reactions
Thermomechanical analysis	TMA	Deformations	Mechanical changes
		Dimensional change	Expansion
Dynamic mechanical analysis	DMA	Moduli	Phase changes, glass transitions, polymer cure
Dielectric thermal analysis	DETA	Electrical	as DMA
Evolved gas analysis	EGA	Gases evolved or reacted	Decompositions
Thermoptometry		Optical	Phase changes, surface reactions, colour changes
Less frequently used methods			
Thermosonimetry	TS	Sound	Mechanical and chemical changes
Thermoluminescence	TL	Light emitted	Oxidation
Thermomagnetometry	TM	Magnetic	Magnetic changes, Curie points

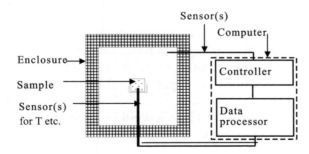

Figure 1 *Schematic of general thermal analysis or calorimetry apparatus*

THE REASONS FOR USING THERMAL AND CALORIMETRIC METHODS

Novice analysts may enquire why yet *another* technique is needed when gas chromatography, molecular and atomic spectrometry and electrochemical analysis plus many other powerful analytical tools are available. The answer might best be given by considering two practical examples.

First, how can you analyse a mixture of processed minerals such as a cement? Although X-ray diffraction might tell you the different minerals present and atomic absorption spectrometry could measure the elements quantitatively, this does not help to analyse how the cement would behave in practice. For this we need to compare the behaviour under conditions of mechanical and thermal stress and the thermoanalytical techniques of TG, DTA and TMA are important tools for doing this.[5,6]

Second, the preparation of new chemicals for new pharmaceutical products, synthetic materials and foods could add to the hazards which workers and customers face. Thermal instability and explosive behaviour can be extremely destructive and costly events. Reaction calorimetry and similar techniques can help to predict the likely behaviour of chemicals when reactions, transport and storage are concerned.[7,8] Physiological behaviour may vary with the nature and form of a drug, and the nature and interconversion of these forms is often studied by thermal and calorimetric methods.

Many analytical techniques require samples in a particular form. For example, gas–liquid chromatography and mass spectrometry need volatile samples and UV–VIS spectrometry usually uses solutions. Therefore, in analysing we destroy the structure of the matrix containing the sample. This has two disadvantages: (i) the behaviour of the sample in its original matrix may be different and (ii) it is time-consuming to alter the form. It is possible to use thermal methods to study the sample "as received". This avoids laborious preparation, does not change the thermal and molecular history of the sample and gives information to the analyst about the real sample and how it would behave in the situation or process where it is actually used.

THE NEED FOR PROPER PRACTICE

Some analytical techniques are sample specific. The "group frequency" bands in an infrared spectrum are largely independent of the method used to obtain the spectrum, whether it is run as a solid KBr disc, a Nujol mull or a solution and whether it is obtained by a dispersive or a Fourier transform instrument. Similarly the titration of an acid with a base should give the same result whether the end-point is detected by an indicator or electrochemically.

This is not always so in the case of thermal methods. The results obtained depend upon the conditions used to prepare the sample, the instrumental parameters selected for the run and the chemical reactions involved. That is not to say that results are not reproducible provided similar conditions are selected. For example, it is possible to compare samples of a polymer to see if their behaviour is "good" or "bad" according to their potential use, but the experimental parameters used for running each sample must be the same.

The useful acronym "SCRAM" (sample–crucible–rate of heating–atmosphere–mass) will enable the analyst to obtain good, reproducible results for most thermal methods provided that the following details are recorded for each run:[9]

The *sample*: A proper chemical description must be given together with the source and pre-treatments. The history of the sample, impurities and dilution with inert material can all affect results.

The *crucible*: The material and shape of the crucible or sample holder is important. Deep crucibles may restrict gas flow more than flat, wide ones, and platinum crucibles catalyse some reactions more than alumina ones. The type of holder or clamping used for thermomechanical methods is equally important. The make and type of instrument used should also be recorded.

The *rate of heating*: This has most important effects. A very slow heating rate will allow the reactions to come closer to equilibrium and there will be less thermal lag in the apparatus. Conversely, high heating rates will give a faster experiment, deviate more from equilibrium and cause greater thermal lag. The parameters of special heating programmes, such as modulated temperature or sample control, must be noted.

The *atmosphere*: Both the transfer of heat, the supply and removal of gaseous reactants and the nature of the reactions which occur, or are prevented, depend on the chemical nature of the atmosphere and its flow. Oxidations will occur well in oxygen, less so in air and not at all in argon. Product removal by a fairly rapid gas flow may prevent reverse reactions occurring.

The *mass of the sample*: A large mass of sample will require more energy, and heat transfer will be determined by sample mass and dimensions. These include the volume, packing, and particle size of the sample. Fine powders react rapidly, lumps more slowly. Large samples may allow the detection of small effects. Comparison of runs should preferably be made using similar sample masses, sizes and shapes.

Specific techniques require the recording of other parameters, for example the load on the sample in thermomechanical analysis. Calorimetric methods, too, require attention to the exact details of each experiment. In the following chapters the principles and practice of thermal analysis and of calorimetry will be described and illustrated with some of the many examples of its use in industry, academic research and testing.

FURTHER READING

An extensive list of reference sources and specialist texts is given in Appendix 1B. Some general texts which introduce thermal analysis and calorimetry for analytical studies are listed here.

General Analytical Chemistry Books with Chapters on Thermal and Calorimetric Methods

G. D. Christian and J. E. O'Reilly, *Instrumental Analysis*, Allyn & Bacon Inc., Boston, 2nd edn., 1986.
F. W. Fifield and D. Kealey, *Principles and Practice of Analytical Chemistry*, Blackwells, Oxford, 5th edn., 2000.
R. Kellner, J-M. Mermet, M. Otto and H. M. Widmer (ed.), *Analytical Chemistry*, Wiley-VCH, Weinheim & Chichester, 1998.
I. M. Kolthoff, P. J. Elving and C. B. Murphy (ed.), *Treatise on Analytical Chemistry*, *Part 1, Theory and Practice (2nd edn.) Vol. 12, Section J*, Wiley, New York, 1983.
D. A. Skoog and J. L. Leary, *Principles of Instrumental Analysis*, Saunders, New York, 4th edn., 1992.
C. L. Wilson, D. W. Wilson (ed.), *Comprehensive Analytical Chemistry*, Elsevier, Amsterdam, 1981–1984, Vol XII, A–D.
J. D. Winefordner (ed.), *Treatise on Analytical Chemistry*, Wiley, New York, 1993, Part 1, Vol. 13.
R. A. Meyers (ed.), *Encyclopedia of Analytical Chemistry*, Wiley, Chichester, 2000.

REFERENCES

1. *Quantities, Units and Symbols*, Royal Society, London, 1971.
2. I. Mills, T. Cvitas, K. Homann, N. Kallay and K. Kuchitsu, *Quantities, Units and Symbols in Physical Chemistry*, IUPAC, Blackwell, Oxford, 1993.

3. R. C. Mackenzie, in *Treatise on Analytical Chemistry*, ed. I. M. Kolthoff, P. J. Elving and C. B. Murphy, Part 1, *Theory and Practice (2nd edn.), Vol. 12, Section J*, Wiley, New York, 1983, pp. 1–16.
4. W. Hemminger and S. M. Sarge, *Handbook of Thermal Analysis and Calorimetry*, ed. M. E. Brown, Elsevier, Amsterdam, 1998, Vol. 1, Ch. 1.
5. *Recommendations for the Testing of High Alumina Cement Concrete by Thermal Techniques*, Thermal Methods Group, London, 1975.
6. H. G. Wiedemann and M. Roessler, in *Proc. 7th ICTA*, ed. B. Miller, Wiley, Chichester, 1982, p. 1318.
7. U. von Stockara and I. Marison, *Thermochim. Acta*, 1991, **193**, 215.
8. M. Angberg, C. Nystrom and S. Cantesson, *Int. J. Pharm.*, 1990, **61**, 67.
9. P. J. Haines, *Thermal Methods of Analysis*, Blackie, Glasgow, 1995.

Chapter 2

Thermogravimetry and Derivative Thermogravimetry

G. R. Heal

*Formerly of Department of Chemistry and Applied Chemistry,
University of Salford, UK*

INTRODUCTION AND DEFINITIONS

Thermogravimetry (TG) is an experimental technique used in a complete evaluation and interpretation of results when it is known as Thermogravimetric Analysis (TGA). The technique has been defined by ICTAC (the International Confederation for Thermal Analysis and Calorimetry) as a technique in which the mass change of a substance is measured as a function of temperature whilst the substance is subjected to a controlled temperature programme.[1] The temperature programme must be taken to include holding the sample at a constant temperature other than ambient, when the mass change is measured against time. Mass loss is only seen if a process occurs where a volatile component is lost. There are, of course, reactions that may take place with no mass loss. These may be detected by the allied techniques of Differential Thermal Analysis (DTA) and Differential Scanning Calorimetry (DSC) which are described in Chapter 3. Results are presented as a plot of mass, m, against temperature, T or time, t. The mass loss then appears as a step. This is shown in Figure 1(A).

The temperature range shown in this plot has been restricted to 400 to 600 °C to show the detail of the step. In a normal experiment the temperature might be run from room temperature to 1000 °C or higher. It should be noted that the shape is sigmoid in nature, that is, although most mass loss occurs around one temperature, where the line is steepest, some

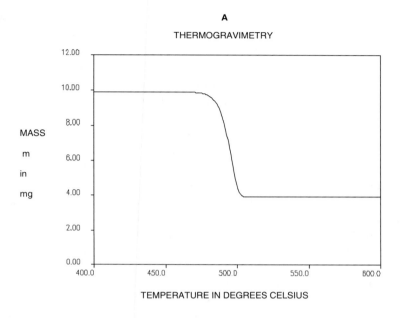

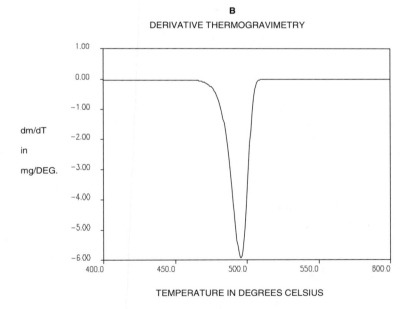

Figure 1 *Typical thermogravimetry results, (A) TG curve, (B) DTG curve*

reaction starts well before the main reaction temperature. Similarly there is still some residual mass loss well after the main reaction.

An alternative presentation of results is to take the derivative of the original experimental curve to give dm/dt, or rate of mass loss against time, and to plot that against temperature, T or time, t. Alternatively the derivative may be against temperature T giving dm/dT. The production of such curves is called Derivative Thermogravimetry (DTG). Such a curve is shown in Figure 1(B); the spread of the reaction over a wide temperature range appears here as a relatively broad peak. The DTG curve is of assistance if there are overlapping reactions. Double peaks or a shoulder on a main peak appear in these cases. Slow reactions, with other fast reactions superimposed, then appear as gradient changes in the DTG curve. The area under the DTG peak is proportional to the mass loss, so relative mass losses may be compared. Measurements of just relative peak heights may suffice for some purposes. The position of the peak may not be indicative of any characteristic point in the mechanism of the reaction, only where mass loss is fastest. However, it may be used, if all that is required is to use the peak as a "finger print" of the presence of a substance in a mixture, *e.g.* a particular mineral in a rock or soil sample.

INSTRUMENTATION

Balance

In the essential form of the apparatus, the substance is placed in a small inert crucible, which is attached to a microbalance and has a furnace positioned around the sample. The furnace may be positioned in several places relative to the balance. This is shown in Figure 2.

The furnace may be above (C), below (A) or around the side arm of the balance (B). A remote coupling system (D) uses magnetic coupling and ensures that the atmosphere around the sample is completely separated from the balance mechanism. The spring balance (E) is a historical version, not well suited to a recording system. The arrangement in the last system (F) has a twin furnace and crucible system to reduce buoyancy effects. The balance is often mounted in a glass envelope, sometimes a metal one. Specialised high-pressure systems use a stainless steel construction. Several types of balance have been used in the past, such as pivoted beam, cantilever beam, and torsion. The modern microbalance has a rotating pivot, as used in a galvanometer, and is controlled electronically using a zero detection device, usually a light and photocell and a magnet and moving coil system to restore balance. The control system varies the current passed through the coil to attempt to keep the beam of

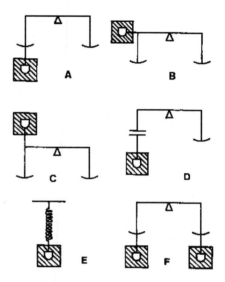

Figure 2 *Layouts for balance and furnace*

the balance in the zero position. This is known as a null deflection system and has the advantage that it keeps the sample in the same position in the furnace throughout the run. Dunn and Sharp[2] have reviewed the accuracy and precision of mass measurements. Early apparatus used large samples of one gram or more, but the modern tendency is to use 10–100 mg, and sometimes only 1 mg. The advantage of gram samples is that sufficient residue is left at the end of an experiment for further tests, such as surface chemical, to be carried out on it. The disadvantage is that the sample will not be at a uniform temperature at any time. Therefore different parts of it will decompose at different temperatures and different rates. There is a lower limit in sample size because to read the sample mass to sufficient precision would require the microbalance to read mass to a fraction of a microgram. The electronic control system introduces random fluctuations at this level, and the balance bench on which the balance stands will introduce vibrations, which are also transmitted to the mass record.

The system of balance plus furnace is called a thermobalance and a typical example is shown in detail in Figure 3.

Modern commercial systems will have a built-in computer system, usually to control the furnace programming and to record and process results.

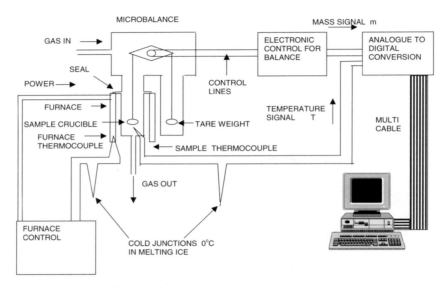

Figure 3 *Schematic diagram of a typical thermobalance system*

Furnace

Furnaces, intended to work up to 1100 °C, use resistive alloy wire or ribbon such as Kanthal or Nichrome, wound on a ceramic or silica tube. For higher temperatures, reaching 1600 °C, platinum or platinum/rhodium alloy is used. The winding is coated with furnace cement to hold the wire in place firmly, because it expands greatly in length during heating. The tube is then mounted in a metal container and packed with insulation material. The modern tendency is to use smaller furnaces and insulating them sufficiently would be difficult, so they have a cooling water jacket around the outside to keep the outer wall at a low temperature. The temperature of a tube furnace varies along its length, and for thermogravimetry, it is good practice to have a long constant temperature zone in the centre. For this reason the winding is made non-uniformly, with turns packed more closely away from the centre but wider near the centre. A furnace control is used to programme the temperature of the sample. Commercial systems available can cool samples to as low as − 160 °C, using liquid nitrogen, or heat up to 1600 °C. The programmed temperature regime is commonly linear, with rates from fractions of a degree to 100 °C min^{-1}. Modern equipment is capable of cooling at a controlled rate as well as heating. It can often offer a more complicated heating regime such as holding at a fixed temperature for a programmed time, then resuming heating.

The furnace is capable of being moved away from the balance case to allow access to the sample. A sliding support allows it to move up, down or sideways as required by the particular design. In many cases a rubber "O" ring produces a gas tight seal between the furnace and the balance case.

Atmosphere Control

The simplest TG experiment would be to heat the sample in static air. However, the sample may react with air in oxidising or burning. Usually an inert gas such as nitrogen or argon is used. In some cases, a deliberately chosen reactive gas is used. This could be hydrogen used to reduce an oxide to metal or carbon dioxide, which affects the decomposition of a metal carbonate. A flowing purge gas is almost always used. This is fed over the balance mechanism first, then around the sample and then out to waste. As well as mass loss by decomposition, thermogravimetry may be used to follow mass gain by reaction with, and uptake of, the purge gas. Also, physical processes such as evaporation of a liquid, sublimation of a solid and desorption of a gas from the surface of a solid may be followed. Most thermobalances will also operate under vacuum as long as all joints are efficiently sealed.

Crucibles

Crucibles are made of various materials. The best ones are made of platinum. These are inert with respect to most gases and molten inorganic materials, and only melt at 1769 °C. If exposed to hydrogen they do chemisorb hydrogen, which might appear as a spurious weight gain. They may also be cleaned in strong acid without any reaction. Unfortunately they are also expensive. They are made of thin platinum to keep the mass low so that they have low heat capacity and follow the furnace temperature without any temperature lag. They must be handled very carefully so as not to squeeze them and distort the shape. Alternative materials are other metals, fused alumina, silica or ceramics. Other metals, such as nickel, may be cheaper, but are less inert. They must never be heated to high temperature in an oxidising atmosphere, such as air. Even in inert gases, such as nitrogen from a cylinder, there are traces of oxygen, and reactive metals may not last over repeated use. Ceramic materials can be inert towards oxygen because they are oxides. However, if the material analysed goes through a molten state, it tends to sink into the solid crucible and is very hard to remove by cleaning.

Thermocouples

The temperature in the system is measured by thermocouples. These consist of two different metals fused into a junction or bead. The junction produces a fixed, standard EMF across the junction, varying with temperature. Strictly a thermocouple system should consist of two such junctions. One is for measuring the sample temperature and this is joined to a second couple held at a constant reference temperature, usually at 0°C in melting ice. In modern equipment a reference EMF is provided electronically.

The metals used are carefully chosen to give a very reproducible, accurate EMF at the junction and preferably as high as possible a value of EMF. They are highly refined pure elements or alloys of more than one element. The commonest thermojunction is probably platinum *versus* platinum alloyed with 13% rhodium. This system has the advantages of high melting point and inertness to samples and purge or product gases. Other metals have been used, but these do not stand heating to high temperatures too often and are better restricted to lower temperatures, *e.g.* only up to 700°C.

One slight disadvantage of platinum, both for crucibles and thermocouples, appears if the temperature is taken above 1000°C, for instance in glass-making studies, where up to 1600°C may be used. At these temperatures there is a tendency for two pieces of platinum to weld together. This might lead to a crucible welding to a thermocouple or the crucible to the hang-down wires or cradle. If this proves to be a problem, then suitable separators must be used. This is often in the form of a thin ceramic plate between the two pieces of platinum. A second problem appears if a second metal is in contact with the platinum. The second metal will have the tendency to dissolve, or alloy, into the platinum, causing holes to appear in the platinum in the worst cases. The remedy is again to separate the metals by an intermediate material. A third cause of difficulty is that platinum is a well-known catalyst and may dramatically increase the rate of any reaction taking place in the crucible.

Temperature Control

As well as the thermocouple system for measurement, a second, entirely separate, thermocouple system is provided to sense the furnace temperature and is connected to the furnace control circuits. The same thermocouple is not used for the two purposes because the criteria for them are very different. The measuring couple has to be positioned as near to the sample as possible. Sometimes this is just below the sample (see

Figure 3), as near as possible, but not quite touching. In other cases the wires for the two halves of the couple are run down the balance support wires and the junction bead actually touches the crucible. The electrical leads are then arranged in such a way so as not to affect the movement or position of the balance.

On the other hand the furnace thermocouple has to be able to respond rapidly to furnace temperature. If there is a lag in time between furnace power being turned up and temperature rise being detected, then the system will tend to go into temperature swings instead of a steady linear rise. For this reason the furnace measuring couple is positioned as near as possible to the source of heat, which is the resistance wire winding. The couple has to be electrically insulated from the winding, but is at least embedded in the furnace cement coating on the furnace.

Data Collection

The control unit often has switches to control mass ranges and to zero the system when an empty crucible is used. There may also be a damping control to dampen swinging of the balance beam. If vibration sets a beam swinging, these swings may continue for a long time before dying away due to the viscosity of the gas in the balance case and will be recorded superimposed on the results. An electronic damping device removes this effect. Since most thermal reactions in the solid state are slow, a high degree of damping may be used. That is, a slow response of the beam to mass loss is not a problem. However, on some occasions faster reactions, perhaps explosive types, may be studied and in these cases the damping has to be turned down or the balance mechanism will miss fast mass changes. A consequence of this is that the trace will become "noisy" and a trade-off between the conflicting requirements has to be made. The output from the control unit is a mV signal representing mass.

Originally data was recorded on chart recorders, but now modern apparatus uses a computer to record the two channels. The analogue signals corresponding to mass and temperature are digitised and the mass and temperature readings are then presented as a graph on the screen and also stored on a floppy or hard disc. The readings may be recalled at a later time to compare with newer results. The computer may be programmed to automatically convert EMF into temperature using a polynomial equation.

A typical commercial system is also shown diagrammatically in Figure 4(A), with the furnace shown in more detail in Figure 4(B).

In this case the sample thermocouple is placed above the sample crucible. There are two purge gas systems. One flow passes over the

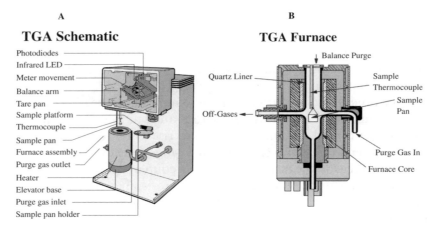

A

TGA Schematic

Photodiodes
Infrared LED
Meter movement
Balance arm
Tare pan
Sample platform
Thermocouple
Sample pan
Furnace assembly
Purge gas outlet
Heater
Elevator base
Purge gas inlet
Sample pan holder

B

TGA Furnace

Balance Purge
Quartz Liner
Sample Thermocouple
Sample Pan
Off-Gases
Purge Gas In
Furnace Core

Figure 4   *A typical commercial thermobalance*
(With acknowledgement to T A Instruments Ltd. Leatherhead, UK)

balance mechanism and then down through the sample area. This keeps hot, maybe corrosive, product gases away from the balance mechanism. A second purge passes across the decomposing sample and the off-gases taken from the left may be led away for analysis. This treatment is mentioned below.

Isothermal Experiments

In many TG experiments, the temperature of the furnace is raised at a constant rate. This type of experiment is referred to as non-isothermal, scanning or rising temperature. An alternative experimental technique is available, and is often used in kinetic studies. Instead of raising the temperature at a constant rate, the temperature is held constant and the mass loss (or mass gain) observed at this fixed temperature. The results are then presented as mass loss against time, t. In practice the sample has to be placed on the thermobalance and the furnace at first left away from the sample. The furnace is then run up to the required temperature and left to stabilise. When the furnace temperature is constant at the required value, the furnace has to be moved quickly around the sample. There are a number of difficulties with this technique. The sample, crucible, thermocouple and cradle have to move rapidly from room temperature to the experimental temperature. They all have a finite thermal capacity, so cannot heat instantaneously. There is a thermal lag while the sample temperature rises. The first part of this rise does not matter, because the reaction being studied will not occur rapidly at lower temperatures. However, as the reaction temperature is approached, some reaction will start at temperatures below the chosen value. At higher set temperatures an appreciable amount of reaction may occur at the "wrong" temperature. This leads to doubt about where the zero for time for the main reaction should be set. It is for this reason that small crucibles and thin wires to hold the crucible are used. Also, a lower mass of sample should be used in this type of study, and 1 mg samples are common. If the furnace is small, then moving it suddenly round a cold sample system causes a dip in the furnace temperature, and the control system takes a finite time to restore the temperature. Another factor is that the flow of purge gas will be upset as the furnace is moved to close the balance case. While the furnace is open the gas flows freely into the atmosphere. When the system is closed the gas has to follow a more tortuous path to escape through tubing. This may cause a drop in flow and affect the zero of the balance.

Calibration for Mass and Temperature

Before a thermobalance is used it should be checked for calibration. The mass reading is relatively easy to check in the same way as for any analytical balance. The balance is first zeroed and then a standard weight, usually in the mg range, is added. If the mass reading is wrong, there are zero adjustments in the control system. The manufacturer's engineer carries out this type of calibration on a routine basis. The measuring thermocouple may be accurate, but may not quite be at the position of the sample, so there is a slight lag between them. One method of checking temperature readings is to carry out decompositions of known samples. A number of standard materials have been suggested for this purpose.[3] The difficulty is that the reaction will not be at a single temperature but spread out over a range of temperatures. A better method is to make use of the Curie point transition in metals. Certain metals and alloys are ferromagnetic at room temperature. When these materials are heated, at a temperature characteristic for each, the material becomes diamagnetic. This change is still not instantaneous but occupies a much shorter range than for a decomposition reaction. If a magnet is placed just above or below the crucible, the sample experiences a magnetic flux in the same direction as gravity. At low temperatures this causes a pull on the sample and a higher mass is recorded. At the Curie temperature there is a sudden loss in apparent mass. Five metals have been quoted as ICTAC Certified Reference Materials for this purpose (see Table 1) and all should be used to give a full calibration for all ranges of temperature. These materials are often supplied by, or are available from, a manufacturer.

On some instruments the sample may be observed visually. It is then possible to place pieces of solid in the crucible and to observe them

Table 1 *Curie temperatures for ICTAC Certified Reference Materials for Thermogravimetry, GM761*

Metal	Temperature of transition/°C	Mean temperature/°C	Standard deviation/°C
Permanorm 3	242–263	253.3	5.3
Nickel	343–360	351.4	4.8
Mumetal	363–392	377.4	6.3
Permanorm 5	435–463	451.1	6.7
Trafoperm	728–767	749.5	10.9

melting directly. It is necessary to stop just below the melting point and to increment the temperature in small steps. All of these measurements result in a calibration graph, or table, for measured thermocouple temperature *versus* the actual temperature from the calibrating sample.

A blank run should be carried out on a balance before it is used for samples. A run with an empty crucible will check the amount of "noise" (random fluctuation) and base line stability (drift). Whatever gas surrounds the sample and crucible it will decrease in density by a large amount as the temperature rises. The sample/crucible/cradle system has a finite volume and displaces this volume of gas, causing a buoyancy effect, which changes with temperature. If the volumes are small and the mass range being used is not too low, the change in buoyancy may not have a measurable effect and may be ignored. To check this, a blank run should be made with the crucible containing an inert sample, *i.e.* one that will show no mass loss during the heating. It should have the same *volume*, not mass, as a sample to be used later. Silica sand is a suitable substance but it should be preheated to high temperature first to make sure there is no water to be lost.

Effect of Experimental Variables

Results Induced by Experimental Conditions

The reaction represented by the results in Figure 1 may be seen to occupy a wide span of temperature. This is because a reaction in the solid state is relatively slow compared to gas or solution reactions due to the fact that molecular movement and collision does not normally control reactions in the solid state. In some cases there may be a diffusion of one or more of the reacting species through the solid lattice, if temperature is high enough, but this is bound to be slow. The rate of a reaction may also be controlled by diffusion of a product gas or by movement of a reacting interface through the solid, reaction being caused by strain of bonds at the interface between the reactant and product solids. In other cases the rate of reaction is thought to be controlled by the rate of transfer of heat to or from the reacting interface. Because of this spread of reaction over time and the fact that temperature is always rising with respect to time, the reaction appears to cover a spread of temperature. For this reason, a careful definition of "decomposition temperature" has to be made.

Figure 5 shows a typical mass loss in a decomposition experiment. The obvious definition would seem to be where the mass loss is steepest, which corresponds to the peak temperature T_p in the DTG plot. However, this is merely the point where reaction is fastest and does not represent the start of reaction, *e.g.* where bonds in the compound begin to break. The position of T_p will depend upon the sample size, packing, and heat flow properties. The point T_i is the initial temperature or onset temperature, but is not easy to identify and depends on the sensitivity of the balance and the amount of drift or "noise" seen. There may be traces of impurities, which decompose or promote some decomposition ahead of the main reaction. A better definition of start of reaction is the extrapolated onset temperature T_e. This requires drawing of tangents to the curve at the horizontal baseline and the steepest part of the curve and marking their intersection. For a reaction that starts very slowly and only speeds up later, T_e and T_i will be very different and a more satisfactory point would be shown as temperature where the fraction reacted α is equal to 0.05, *i.e.* $T_{0.05}$. Another definition of reaction temperature, important in kinetic studies, is when the reaction is half over, that is, when the fraction reacted

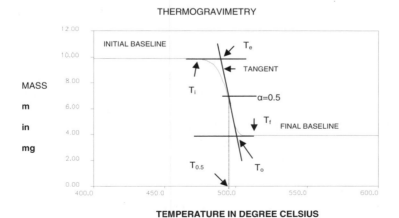

Figure 5   *Definition of the decomposition temperature on a TG curve*

$\alpha = 0.5$; this is $T_{0.5}$. To show the complete temperature range for reaction, two more temperature values may be added. These are T_f the final temperature (again difficult to pick out accurately) and T_o the extrapolated offset temperature.

Although the sample may be decomposing at a temperature which is characteristic of the compound, the shape of the decomposition curve will be affected by many factors. The particle size may control the rate of diffusion of a reactant or product. Sometimes large sized particles may split so violently that pieces jump out of the crucible, causing a spurious mass loss record. The rate of flow of heat may be controlled by the type and size of the crucible and by the particle size and degree of packing and mass of the sample. Large samples tend to have a temperature gradient through them leading to early decomposition of the outer part and delayed decomposition of the centre part. A thin film of powder gives the lowest commencing and finishing temperature, followed by a thick film of a fine powder, and then by compressed pellets of the material at the highest temperature. For these reasons, small sample sizes are recommended, although it has been shown that grinding may alter the crystalline structure and the course of the decomposition. It should be remembered that all reactions involve an enthalpy change and the heat released will heat or cool the sample slightly relative to the programmed furnace temperature. Many of the solids investigated will be poor conductors of heat, which will exacerbate the effect. This is another good reason for small samples, keeping a thin layer of sample as near as possible to the crucible temperature. The relative sizes of the sample crucible and furnace will also change the heat flow. The material of the crucible will have an effect, metals like platinum being good conductors of heat, while alumina is a much poorer conductor.

The reaction may be reversible with respect to the product gas, so a varying pressure of product gas around the sample would affect the kinetics of the decomposition and thus the shape of the decomposition curves. The shape of a container may have a slight effect. Shallow or deep containers may cause the atmosphere at the sample surface to differ if the gas evolved is not the same as the purge gas. If the decomposition product gas does build up in deep crucibles, or deliberately closed crucibles, this is referred to as a "self-generated" atmosphere. If the intention is to carry out decomposition into pure product gas, deliberately flowing over the sample, to produce an equilibrium effect, then that is a different experiment to the one with an inert purge gas. The effect of changing the purge gas may be very large. This is illustrated below in the example on oxysalt decomposition, where the product may react with oxygen in air, but an entirely different product is left if nitrogen is used. In the case of the decomposition of anhydrous calcium, strontium and barium oxalates the gas evolved is CO and the carbonate is formed. This is true whether air of nitrogen is used. However, the CO is oxidised by oxygen in the air as a surface reaction on the surface of the oxalate–carbonate mixture. This reaction is exothermic and produces local heating of the remaining reactant, causing a slight acceleration of its decomposition. No such effect appears for a nitrogen atmosphere, so the decomposition traces differ slightly when different atmospheres are used.

The rate of heating will have a major effect on the result, because, at a high heating rate, the temperature recorded moves to higher values while the reaction is taking place slowly, *i.e.* the reaction appears to be at a higher temperature. This effect is illustrated in Figure 11 below.

In some high temperature experiments a solid present may become volatile and sublime out of the crucible. This will have to be accounted for in the description of the mechanisms taking place. However, in unfortunate cases, some of the solid may redeposit on the support rod or wire for the crucible in slightly cooler regions. This will cause an apparent mass gain, partially cancelling the mass loss, and invalidating the results. The support system should be examined frequently for this effect and any deposits cleaned off thoroughly.

In summary, experiments should be carried out with high thermal capacity furnaces and with small, lightweight crucibles. The rate of heating should be low so that reaction can take place over a narrow range of temperature. The sample size should be as small as possible and should be spread thinly on the base of the crucible. A difficulty arising is that if a sample is coarse-grained then only a few particles may be included and may not be representative of the sample if, say, minerals or concrete are being tested. In this case the sample should be ground in a pestle and mortar to a fine powder and well mixed before sampling.

From the above remarks it may be seen that many factors affect the observed results. This means that although thermogravimetry can identify a substance from its decomposition temperature it should not be thought of as a "finger print" method like spectroscopy. In that technique a peak will always be at the same position in the spectrum, regardless of the make of the instrument or size of sample.

A consequence of this is that all experimental variables should be reported thoroughly to allow for variations between instrument and experimental conditions. A rigid set of rules for making reports has been drawn up and should always be adhered to. It is important to ensure that someone reading the report would be able to repeat the experiment exactly and get, as near as possible, the same result.

REPORTING THERMOGRAVIMETRY RESULTS

Recommendations for reporting thermal analysis results, including thermogravimetry results, have been made by the Standardisation Committee of ICTAC and appear in standards such as ASTM E 472 (1991). Not all of the points listed may be known, if commercial equipment is used, but as much as possible should be reported.

A. Properties of the Sample

(1) Identification of all substances (with %, if an inert diluent is used).
(2) Give the source of all materials used (if commercially obtained, give manufacturer, grade and purity).
(3) List the history of the sample before it was sampled (this might be just "from the bottle", but might include grinding, sieving, pre-drying in an oven, surface modification, *etc.*).
(4) Give physical properties if known (particle size, surface area or porosity).

B. Experimental Conditions

(1) State the apparatus used (manufacturer and model name or number and if modifications have been made).
(2) Describe the thermal treatment during the run (initial temperature, final temperature, rate of heating if linear or full details if not linear).
(3) Identify the sample atmosphere (flow rate, pressure, composition and purity). Remember that cylinders of so called pure inert gas, such as "white spot" nitrogen, contain traces of oxygen (try holding a carbon sample in such a nitrogen flow at $1000\,^\circ C$ and observe the mass loss!). Also state if static (zero flow), self-generated (not to be encouraged) or dynamic (flowing) atmosphere is used.
(4) State the dimension, geometry and material of the sample holder (crucible). Also give the method of loading, *e.g.* tipped into the crucible on the balance or weighed out on a separate balance and tapped down on a hard surface.
(5) Give the sample mass.

C. Data Acquisition and Manipulation Methods

Note: much of this may be covered by just quoting the apparatus used.

using a pair of curved tweezers with pointed ends. It is very common for beginners to drop the crucible into the furnace and often the sample powder as well. For this reason, place a plate over the furnace entrance at all times when the crucible is being removed or replaced. A piece of sheet aluminium is suitable and can be used also when heating the furnace for isothermal experiments. If a mistake is made and the crucible *is* dropped into the furnace, under no circumstances should the operator "fish" inside the furnace using the tweezers to retrieve the crucible. This is because the tweezers could damage a thin wire thermocouple mounted so that it is just below the sample, when the furnace is raised. Instead the furnace must be disassembled and inverted to retrieve the crucible and/or tip out powder that has been spilt.

The crucible must be cleaned. Many crucibles are made of platinum, which is inert to acid and heating up to $1769\,°C$; but if alumina, ceramic or another metal is being used, the instructions given may have to be modified so as not to destroy or oxidise the crucible. To clean, place the crucible in a small beaker and soak it in moderately concentrated nitric acid. This should remove most inorganic substances, because all metal nitrates are soluble. Wash thoroughly with water. Now place the crucible on a gauze and triangle and place a hot Bunsen burner under it. In theory the crucible should be heated to at least the temperature to be used in the run. In practice, glowing white-hot is sufficient. Allow the crucible to cool and then transfer it empty to the cradle or holder on the balance. While the crucible was being cleaned your fingers could handle it, but after heating it should only be touched by the tweezers, the tips of which have also been flamed. Remember! fingers put fingerprints, which consist of grease, onto the crucible. These will evaporate off and be recorded as a mass loss during the run.

Zero Setting

After the empty crucible is on the balance, move the furnace back into position around the crucible and start the flow of the purge gas that is used to remove product gasses. The flow rate is usually about $10\,cm^3$ min^{-1}. Do not use a fast flow because it may influence the balance, causing it to sway and so give a fluctuating mass record. Change the balance control to the released position and set a suitable mass range, typically $10\,mg$ if $10\,mg$ samples are used. Carry out the procedure in the computer system to store the zero mass reading.

Adding the Sample

Arrest the balance again and lower the furnace. The calcium oxalate, which is a white powder, may now be added. This may be done with the crucible on the balance using a micro spatula and a steady hand!

A plate over the furnace entrance is essential or, alternatively, the crucible may be removed and the sample added away from the balance. The exact mass used is not usually critical, so guess work may be used, the exact mass being found when the crucible is back on the balance. If the mass reading is critical, then a separate standard laboratory balance may be used in the normal way to weigh out the sample. Replace the crucible on the balance and raise the furnace. Switch the balance to the released position and leave for the gas flow to be established and the sample to stop swaying.

Starting the Run

If the furnace temperature has a dial control, turn it back to room temperature, *i.e.* 20 °C. On other systems set the computer control to a start temperature. Set the rate of heating to the required value. This is typically $10 °C$ min^{-1}, but might be 1, 20 or $50 °C$ min^{-1} for certain experiments. There should be some means of setting the maximum temperature reached, perhaps on the dial that indicates rising temperature or otherwise on the control panel. Set this to 1000 °C for the experiment described. The control system may give alternative control programmes such as controlled heating followed controlled cooling. In the present experiment, set the control to heat to the maximum temperature and switch off. If the furnace has a cooling jacket, turn on the water flow and adjust the flow rate; it only needs a slow flow. Next switch on the data logging system. When the mass reading is steady, turn on the temperature control to heat. The run should now be automatic.

Ending the Run

When the maximum temperature is reached the furnace control should switch off. If cooling water was used, leave it running until the furnace cools to near room temperature, then turn it off. Turn off the furnace control, balance control, water supply and gas purge. The computer collecting the data should show the mass loss on the screen against sample temperature. The plot may be printed and the data may be stored.

Example Experiment

Details of the experimental conditions for the typical experiment above.

Apparatus used – any suitable type of thermobalance.

Sample – calcium oxalate monohydrate powder, Hopkin & Williams, Standard Grade.

Sample was used untreated.

Scan conditions – 20 °C to 1000 °C at a linear rate of 10 °C min^{-1}.

Purge gas – nitrogen, "white spot", passed through drying tube; flow 20 cm^3 min^{-1}.

Furnace cooling water – 5 cm^3 min^{-1}.

Crucible – platinum, cylindrical; diameter 0.5 cm, height 0.5 cm.

Sample mass – 8.404 mg; tamped.

If there is no built-in differentiating procedure, applying the Savitzky–Golay[4] method, using a 15-point smoothing/differentiation routine, based on a quadratic fit, produces the DTG result.

Results

Figure 6(A) shows the mass loss as the temperature is taken up to 1000 °C and may be seen to consist of three stages or steps. Figure 6(B) is the DTG curve, that is the derivative form of Figure 6(A). This helps to pinpoint the point of maximum weight loss. It should be noted that although the TG curve is relatively smooth the DTG curve has many fluctuations. This is caused by minor variations in the TG curve. It is a characteristic of the differentiation process that "noise" on an experimental curve is accentuated. The first decomposition is widely separated from the second stage. However, the second stage only just finishes before the third stage starts. In less favourable decompositions two such steps may overlap with no horizontal part between the stages on the TG curve and no return to the baseline on the DTG plot. In some of the worst cases a DTG peak will appear as a small shoulder on the side of a larger peak. In these cases the experiment should be re-run using a much slower rate of heating to try to allow one reaction to finish before the next starts.

The first stage, which takes place between 100 and 200 °C, is likely to be loss of water of crystallisation since the substance is known to be a salt hydrate. The decomposition of the anhydrous material takes place in two steps. Some oxalates are known to decompose through an intermediate carbonate, so this may account for the two steps. The final product of decomposition is usually the commonest oxide (*i.e.* the most stable oxidation state), but sometimes the metal is formed in an inert atmosphere. A possible reaction scheme is given, together with the molecular masses of the compounds:

1st stage $\qquad\qquad Ca(COO)_2 \cdot H_2O = Ca(COO)_2 + H_2O$ (2)
$\qquad\qquad\qquad\quad$ 146.115 $\qquad\qquad$ 128.100 $\qquad$ 18.015

2nd stage $\qquad\qquad\quad Ca(COO)_2 = CaCO_3 + CO$ (3)
$\qquad\qquad\qquad\qquad$ 128.100 $\qquad$ 100.089 $\;$ 28.011

3rd stage $\qquad\qquad\qquad CaCO_3 = CaO + CO_2$ (4)
$\qquad\qquad\qquad\qquad$ 100.089 $\quad$ 56.079 $\;$ 44.010

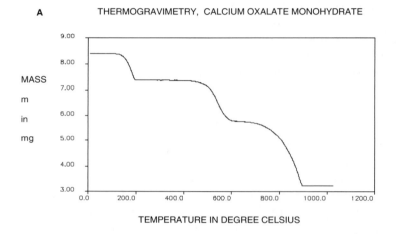

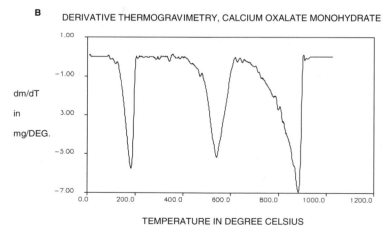

Figure 6 *Results of a typical TG experiment – calcium oxalate monohydrate, (A) TG curve, (B) DTG curve*

The theoretical percentage mass losses may now be calculated. Care should be taken to ensure that the ratios are taken to the original starting material, *i.e.* the hydrate, not as a percentage of the starting compound in each reaction step. The masses of the sample at the start and after each step were read. This could be carried out from a printout of Figure 6(A), or, more accurately, from a listing of the data. The mass readings were 8.404, 7.391, 5.744 and 3.238. From these the experimental percentage mass losses were calculated. These results are shown in tabular form below:

	Theoretical loss (%)	Experimental loss (%)	Peak Temperature T_p /°C
1st stage	12.33	12.05	182.4
2nd stage	19.17	19.60	540.3
3rd stage	30.12	29.82	882.9

Within experimental accuracy, given that the material was not particularly pure, these results confirm the equations suggested above. The peak temperatures were taken from the DTG curve.

Another point worth mentioning is that the loss of water occupies a much smaller temperature range than the other two reactions. Also, the carbonate decomposition shows a very abrupt end, shown by the steep return to the baseline on the DTG plot. This is characteristic of carbonate decompositions.

APPLICATIONS

Oxysalt Decomposition

Amongst the studies of oxysalt decomposition has been one on oxalates.[5] A number of metal oxalates were subjected to a thermogravimetric run in air and nitrogen up to 1000 °C. The object was to see if there were any differences in behaviour between metals, to determine the end product, to see if the atmosphere made any difference and to check the experimental mass losses against the theoretical ones to confirm the reaction mechanism. The shapes of the curves produced fell into 5 types as shown in Figure 7.

The basic shape of curve was type A. In this case two distinct steps were seen: the first was due to water of crystallisation loss and the second due to the oxalate decomposition. Type B showed no distinct end to water loss but the two reactions appeared as one continuous mass loss. Type C was found for samples with no water of crystallisation and showed only

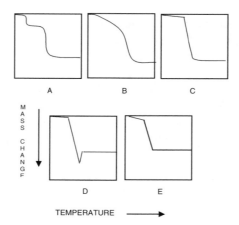

Figure 7 *Shapes of TG curves*

surface moisture loss. However, the two iron samples followed this shape also. Although, from their formulae, they contained water, the end of the water loss could not distinctly be determined, but the oxalate decomposition started sharply and clearly. Type D showed the water loss and the oxalate decomposition showed a "spike", when carried out in air. This was due to the formation of an inert atmosphere of product gas in the crucible, which drove out oxygen for a while, causing the formation of some metal or lower valency oxide. When oxygen diffused back in, the metal or lower oxide oxidised to the normal oxide, causing a mass gain. Type E was like type C but showed a very sharp end to the oxalate decomposition. The results for the various oxalates are shown in Table 2.

It may be seen that a change in atmosphere often causes a change in shape of decomposition curve. Calcium, strontium and barium oxalates, however, behave somewhat differently. Their carbonates are all more stable than their oxalates. For this reason their oxalates decompose to carbonate first and to oxide at a higher temperature. This is shown for calcium oxalate in Figure 6 and discussed above.

Table 3 shows the close agreement of experimental and theoretical mass losses for the reaction mechanisms chosen. In some cases the atmosphere used makes no difference to the final product. In other cases the use of nitrogen changes the product from oxide to metal or higher oxide to lower oxide. In some cases the percentage of water is rather indefinite. This is probably because some salt hydrates tend to lose water to the atmosphere, or the reverse and take up water as a higher hydrate or on the surface. Thus the true formula of the hydrate may not be exactly what is given on the sample bottle. Aluminium oxalate is anomalous. It

Table 2 *Classification of thermogravimetric results into types of shape of the curve*

Oxalate	Decomposition in		Oxalate	Decomposition in		Oxalate	Decomposition in	
	Air	N_2		*Air*	N_2		*Air*	N_2
Copper	D	E	Lead	D	E	Manganese	A	A
Zinc	A	A	Thorium	A	A	Ferrous	C	C
Cadmium	E	E	Antimony	D	E	Ferric	C	–
Aluminium	B	B	Bismuth	D	E	Cobalt	A[a]	A
Tin	E	E	Chromium	B	B	Nickel	A[a]	A

[a] Rapid heating in a limited supply of air can result in a curve of type D.

has water of crystallisation but holds it strongly, and only loses it slowly over a wide temperature range. This process overlaps with the oxalate decomposition and is continuous up to 1000°C, where an indefinite amount of water remains. Thus it is impossible to make percentage loss comparisons.

Polymer Stability and Charcoal Production

Amorphous carbons may be prepared in an infinite variety of structures. Many of these structures are porous, with pores of sizes down to molecular dimensions and very high surface areas, up to several thousand m^2 g^{-1}. These materials are important in surface chemistry as adsorbents in their own right or as supports for other materials such as catalysts. Sometimes the pores are required to be small, so that the material will act as a molecular sieve, and sometimes the pores need to be large to allow reacting gases to diffuse in and out easily. Traditionally such materials are made by carbonising (heating in the absence of air) such precursors as wood, coal or pitch. These starting substances are characterised by being polymeric, and heating drives off volatile components or small molecule substituents from the polymer chains. The materials mentioned are necessarily very diverse in make up, variable in degree of polymerisation and contain large amounts of impurities that are liable to remain in the final charcoal. It was suggested some years ago that better information could be obtained about the carbonisation process if simpler substances were used as the starting point.[6] Tests showed that pure polymers would fit this requirement. They should undergo decomposition in relatively simple steps of reaction and should leave no impurities in the charcoal unless the residue of an initiator is left behind in the polymer and thus

Table 3 *Thermogravimetric analysis of some oxalates*

Oxalate	Decomp. in	Loss % on dehydration		Loss % on decomposition		Loss % calc. on anhydrous		End product[a]
		Found	Calcd.	Found	Calcd.	Found	Calcd.	
$CuC_2O_4 \cdot \frac{1}{2}H_2O$	Air	3.5	5.6	47.9	50.0	47.9	47.5	CuO
	N_2	3.5	5.6	58.6	60.4	57.1	58.0	Cu
$ZnC_2O_4 \cdot 2H_2O$	Air	18.5	19.0	55.3	57.0	45.2	46.9	ZnO
	N_2	18.3	19.0	55.3	57.0	45.3	46.9	ZnO
CdC_2O_4	Air	6.0	0[b]	40.5	35.9	36.7	35.9	CdO
	N_2	6.0	0[b]	48.5	44.0	45.1	44.0	Cd
$Al_2(C_2O_4)_3 \cdot 3\text{–}6H_2O$	Air							$Al_2O_3 \cdot xH_2O$
	N_2							$Al_2O_3 \cdot xH_2O$
SnC_2O_4	Air	0	0	27.1	27.1	27.1	27.1	SnO_2
	N_2	0	0	34.2	34.8	34.2	34.8	SnO
PbC_2O_4	Air	0	0	25.0	24.5	25.0	24.5	PbO
	N_2	0	0	29.2	29.8	29.2	29.8	Pb
$(SbO)_2C_2O_4$	Air	0	0	19.8	19.8	19.8	19.8	$Sb_2O_3{}^c$
	N_2	0	0	27.0	28.6	27.0	28.6	$SbO_2{}^c$
$Bi(C_2O_4)_3 \cdot 4H_2O$	Air	6.0	6.0	37.0	35.7	33.0	31.7	Bi_2O_3
	N_2	6.0	6.0	39.7	40.3	35.8	36.4	$Bi_2O_3{}^d$
$Th(C_2O_4)_2$	Air	8.5	8.1	41.0	40.5	35.5	35.3	ThO_2
	N_2	8.3	8.1	41.1	40.5	35.7	35.3	ThO_2
$Cr_2(C_2O_4)_3 \cdot 6H_2O$	Air	NF		69.5	68.1	–	–	Cr_2O_3
	N_2	NF		70.0	68.1	–	–	Cr_2O_3
$MnC_2O_4 \cdot 2H_2O$	Air	20.1	20.1	57.4	57.4	46.6	46.7	Mn_3O_4
	N_2	20.2	20.1	60.3	60.3	50.3	59.3	MnO
$FeC_2O_4 \cdot 2H_2O$	Air	NF		55.6	55.5	–	–	Fe_2O_3
	N_2	NF		57.0	55.5	–	–	Fe_2O_3
$Fe_2(C_2O_4)_3 \cdot xH_2O$	Air	NF		63.0	–	–	–	Fe_2O_3
	$N_2{}^e$	–		–		–		Fe_2O_3
$CoC_2O_4 \cdot 2H_2O$	Air	21.0	19.7	55.0	56.1	43.0	45.3	Co_3O_4
	N_2	19.7	19.7	67.7	67.8	59.8	59.9	Co
$NiC_2O_4 \cdot 2H_2O$	Air	22.0	19.7	61.5	59.1	50.6	49.1	NiO
	N_2	19.7	19.7	67.8	67.8	59.9	59.9	Ni

[a] Used for the calculation of theoretical mass loss. [b] Sample contained moisture which could be removed by heating at 100°C. [c] Mixture of Sb and SbO. [d] Mixture of Bi and BiO. [e] Not studied. NF = not formed.

contaminates the charcoal. Surface chemical tests on the charcoals show that pore structure and surface area change from one starting polymer to another. Also, conditions of the reaction, such as heating rate, final temperature and physical state of the polymer – such as loose powder or compacted disc – varies the final result. It seems that a study of the carbonisation process may allow a charcoal to be "tailor-made" for a particular adsorptive property.

Some polymers do not carbonise on heating but instead de-polymerise into small chain pieces. For instance polyethylene and polystyrene only yield hydrocarbons of a variety of chain lengths and no residue at all. These are, of course, useless for the desired process.

In one study[6] fifteen polymers were heated from room temperature to final temperatures between 600 and 900 °C on a thermobalance. The sample size was 100 mg in all cases and was placed in a platinum crucible of size about 1 cm diameter by 3 cm high. The purge gas was nitrogen from a "white-spot" cylinder, but was further purified to remove traces of oxygen and dried. The gas flow rate was 30 cm^3 min^{-1}. The heating rate was 10 °C min^{-1}. The residue was cooled in the nitrogen flow and recovered for surface chemical studies. The recording system used two pens to record furnace temperature directly in °C and mass in mg. The crucible stood on a platform above the balance mechanism. A thermocouple was run up the hollow platform support rod, with the bead touching the crucible to measure sample temperature. The EMF of this thermocouple was connected to a separate pen recorder set on a 10 mV range. The sample lagged behind the furnace temperature by about 30 °C. Readings were taken from the charts at 10 to 20 degree intervals and entered into an EXCEL spreadsheet and plotted as a graph. Other details are unknown. Six examples of the resulting graphs are shown in Figure 8.

The mass of material lost as volatile matter was high in cases A, B and C, but much lower for E and F. Heating poly(vinyl alcohol) only left 5% by mass of carbon; not a very efficient way of making a carbon. Poly(vinylidene chloride) only left 25% carbon, but this is known to be an important fine-pored carbon, capable of adsorbing carbon dioxide but not nitrogen, and so is worth studying. The shapes of the decomposition curves are very varied. There is obviously more than one mechanism at work for each sample. The reactions are not well separated but run into one another or overlap. Early reactions are probably necessary steps before a second reaction can start. A simple thermal conductivity gas detector was used to find when a gas was being evolved. Tar was also seen to condense out in a "U" tube at certain points during the decomposition. Graphs A, B and C all show a very steep large mass loss. Gas was shown to be evolved in large quantities during this stage. In A and B a separate

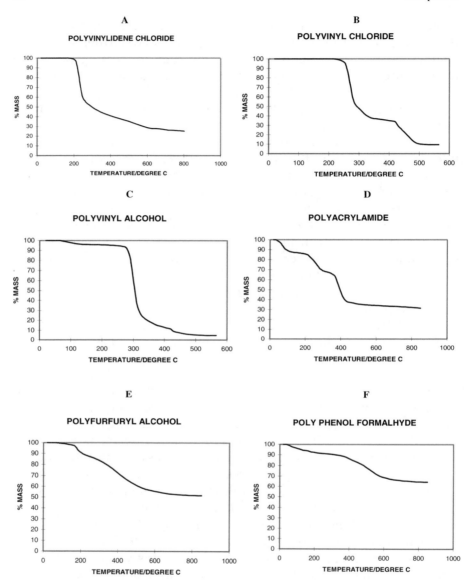

Figure 8   *Examples of TG of polymers during carbonisation*

test showed this to be HCl vapour. This part of the decomposition is due to an "unzipping" reaction. One molecule of a substituent, *e.g.* HCl, splits off and this promotes the removal of the next adjacent HCl molecule in the polymer chain. This effect then runs along the chain rapidly, or "unzips" the molecule. The later stages of decomposition of these three polymers do not show any volatile gases, but tar appears in the condensing tube showing that larger fragments of polymer molecule are being split off. The samples that show mass loss at low temperature, *i.e.* below 200°C, are found to be due to loss of water. The successive steps for polyacryamide, sample D, are loss of water, NH_3 and tar. The last two samples E and F do not appear to show the "unzipping" type of reaction but show steady mass loss at all temperatures with a slight variation in rate at various points. The material evolved seems to be almost entirely tars. These are the materials with a large residue, so it seems that it is "unzipping" reactions that cause the largest mass losses and poor yield of carbon. All samples have a tendency to still show slow mass loss as the temperature goes higher after the main reactions are over. Once the stepwise nature of decomposition has been established, temperatures may be estimated for the end of one process or the start of another. Then runs may be conducted holding temperature fixed at one point and examining the intermediate product or then applying further heating. This allows novel carbons to be designed. This study is further expanded upon in reference 6.

Of interest in studies of polymer degradation are two standard tests, the ASTM D 3850 (1994) test method for rapid thermal degradation of solid electrical insulating materials by thermogravimetric method, and the ASTM D 6370 (1999) standard test method for rubber compositional analysis by thermogravimetry.

Metal Oxidation

This is an example of mass gain instead of the usual mass loss, but may be equally well studied on a thermobalance.[7] Oxidation of metals in gaseous oxygen is an example of a process controlled by transport across a layer of product. Metals other than gold spontaneously convert into the corresponding oxide, even at room temperature, due to the metal being thermodynamically unstable with respect to the oxide. In practice the oxidation ceases long before the metal is consumed. Oxide layers builds up and further oxidation only proceeds by transport through the layer. Often the oxide provides an impenetrable barrier and oxidation virtually ceases. This is sometimes called passivation. This limiting layer is usually only of the order of nm in thickness. It is thought that either metal ions move through the oxide to the surface and are there oxidised or oxide ions move the opposite way to the metal/oxide interface to react. The former is more likely because most metal cations are smaller than oxygen ions. Raising the temperature speeds up the process – so these studies are conducted at an elevated temperature.

Oxidation has been measured using sheets of the metal, placing them on a thermal balance in an oxygen flow, and holding them at an elevated temperature to accelerate the process.[7] Examples are shown in Figure 9.

It might have been thought that oxide build-up would continuously slow the reaction and the curve would show a decreasing slope, eventually becoming horizontal. In fact, for the oxidation of magnesium, shown in Figures 9(A) and (B), a linear rate of oxidation appears, after an initial settling down period. It is postulated that the bulk of the oxide has become porous to gaseous oxygen. Only a thin barrier layer of oxide controls the rate of oxidation. The oxide is thought to crack or re-crystallise to form a porous material and the rate of cracking equals the rate of thickening of the barrier layer. Many metals show changes in rate of reaction at various points. At point B in Figure 9(B) there is a relatively sudden increase in rate. This is termed "breakaway" or "rate transition". It is thought that the oxide layer grows to some critical dimension when stresses cause it to crack away allowing oxygen easier passage to the metal. Note that this happens at 525 °C for magnesium, but not in Figure 9(A) at 500 °C.

Figure 9(C) shows for niobium the increased rate of oxidation as temperature is raised. In the case of the 450 °C curve this also shows no difference for dry and wet oxygen over the relatively short time of 7 h whereas Figure 9(E) shows that, on the longer time scale of 100 h, the reaction curves for dry and wet oxygen diverge greatly. The result in Figure 9(D) is an expansion of the very initial part of Figure 9(E) for 120 min. "Breakaway" behaviour is shown for niobium at 450 °C at three points: Point X on Figure 9(D); points C and D on Figure 9(E). The highest part of Figure 9(E) shows the reaction slowing down because almost all of the metal has been consumed.

Compositional Analysis

Compositional analysis of coal, polymers and rubbers deals with the determination of volatile materials under an inert gas atmosphere and then combustibles under an oxidising atmosphere. The procedure is described in ASTM E 1131 (1998).[8] The thermobalance must be capable of rapid heating rates and the system must allow gas switching to be performed quickly and completely. The quality of the inert atmosphere is most important and must not permit any significant oxidation during the determination of the volatiles. The proximate analysis of coal is shown in Figure 10.[9] Using classical analytical methods, these determinations would take several hours.

Using a sample of about 10 mg, the moisture loss is determined by

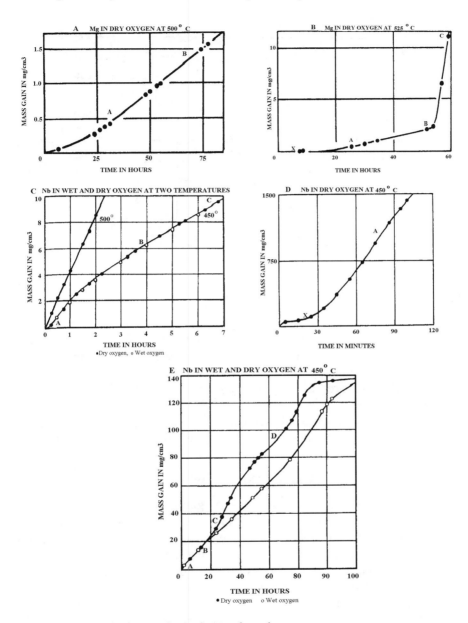

Figure 9 *Typical TG curves for oxidation of metals*

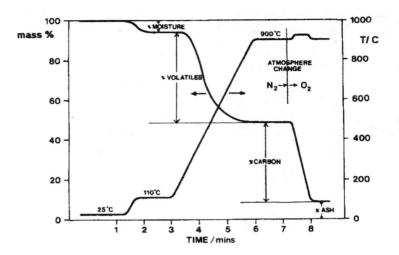

Figure 10   *Proximate analysis of a coal*

heating to 110 °C and the volatiles by heating to 900 °C, both in an atmosphere of pure, oxygen-free nitrogen. The carbon content is then measured by switching to an oxygen atmosphere and the residue measures the ash present.[10,11] The entire analysis takes less than one hour.

Compositional analysis of polymers, rubbers and foods, as well as mixtures of inorganic salts may also be carried out by thermogravimetry.

Glass-making Reactions

There are a large number of types of glass made according to the properties required. As well as the main raw materials there are many impurities and substances added in small quantities to produce special effects. A typical clear, flat, soda–lime–silica glass would be made from the following materials (with the proportions given in grams): sand (silica) 74.0; dolomite (calcium magnesium carbonate) 16.5; soda ash (sodium carbonate) 22.0; limestone (calcium carbonate) 5.5; salt cake (sodium sulfate) 0.8, and carbon 0.02. These are processed to suitable particle size, and mixed. This mixture is called "batch" and is fed continuously into a furnace at one end and glass flows out at the other. The temperature along the furnace changes, but up to 1550 °C is reached. The final product could be said to a mixture of Na_2O, CaO and SiO_2, plus impurities. This is an oversimplification and the reactions involved must be very complex.

It would be impossible to study all of the reactions simultaneously by thermal analysis. Instead, the technique of investigating two, or perhaps more, pure substances reacting together has been used. The results are then taken as representative of one of the many processes occurring and used to forecast what happens in the overall "batch" reaction.

The sand used is a mixture of silica particle sizes and, since much of the time to form glass is in dissolving this silica, it is of interest to study silica's reaction with sodium carbonate.[12] In reference 12 silica was sieved to obtain specific particle size ranges and the sized samples examined separately. For example, 2 mol of the 425 μm silica was mixed with 1 mol of sodium carbonate and heated at three heating rates from 5 to 30°C min^{-1}. The result is shown in Figure 11.

In the case of a simple, single reaction, changing the heating rate should make no difference to the final mass loss, only the time taken and the temperature where final mass loss is achieved will change. In the present case the heating rate change is seen to alter the final mass loss. This may be immediately taken to mean that a complex set of reactions must be taking place and that the change in heating rate is favouring one reaction with respect to another.

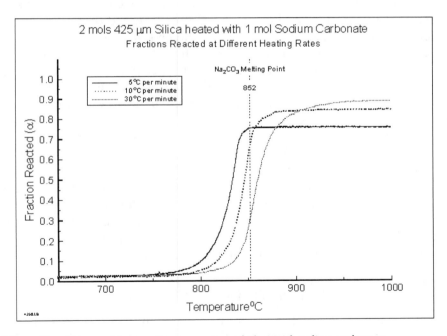

Figure 11 *Glass-making reactions – reaction of silica with sodium carbonate at various heating rates*

The melting point of Na_2CO_3 is 852°C, which is indicated on the Figure 11. The mass loss is due to the evolution of CO_2 as sodium silicates are formed. Both sodium metasilicate (Na_2SiO_3) and sodium disilicate ($Na_2Si_2O_5$) are formed at various points in the reaction. It is obvious from the figure that the process is not simply melting of the carbonate followed by dissolution and reaction of the silica because the mass loss starts before the carbonate melting temperature. Indeed, at 5°C min^{-1} almost all of the reaction is over before the carbonate melts. It has been found that eutectic melts occur for mixtures of compounds. A metasilicate/disilicate eutectic melts at 790°C and a silica/disilicate eutectic melts at 840°C. Thermodynamic calculations suggest it is the disilicate that is formed first at temperatures as low as 400°C. This is the mechanism at the points of contact between the particles of the two solids; but a solid/solid reaction is very slow. Eutectic melting is then thought to occur and more reaction can then take place at a faster rate as a solution reaction. Figure 11 appears to show that reaction is delayed to higher temperatures if the rate of heating is faster. As explained above this is only because the reaction is slow and the sample temperature is taken higher more quickly at the faster heating rates and the reaction takes time to catch up. The ratio of substances chosen was 2 of SiO_2 to 1 of Na_2CO_3 because this is correct for the formation of disilicate. It must be noted that if this exact reaction took place the final value of α would be 1. In fact $\alpha = 1$ was never reached and the value obtained varied with heating rate. This is thought to be because some of the carbonate does not react and just dissolves in the melt without reaction, together with unreacted silica. An XRD pattern showed no crystalline compounds present in the product, only glassy material. The differences in the ending value may be due to varying amounts of unreacted products dissolved at the end, but may also be due to amounts of sodium metasilicate produced by a reaction, which is possible at the higher temperatures.

Further experiments were also carried out using various silica particle sizes with only one rate of heating. Details may be found in reference 12. The reactions are thus seen to be complex, and ancillary information, such as the analysis of the residue material or the volatile material given off and condensed out, helps in formulating a reaction scheme.

Kinetics of Solid-state Reactions

This topic is a large one and cannot be covered fully in this book. There are very many research papers on the subject, and also reviews of the topic,[13-17] which should be consulted. An outline only of the techniques and problems will be given here.

The reacting species are normally not free to move through the mass of solid and so cannot be thought to be at a particular even concentration within the solid. Even if a diffusion process controls the mechanism, diffusion is difficult and slow in the solid state, controlled by concentration gradients. The reacting molecules do not move freely and collide at a rate controlled by the thermal energy of the system as happens in gases and liquids. Indeed a reaction interface may move through the mass with pure reactant ahead of it and pure product behind. Thus concentration is meaningless and is replaced by fraction reacted, represented by α. The mechanisms that have been postulated are varied. Nucleation is a common starting point. This involves the breaking of some bond in a molecule within the structure, followed by rearrangement and the release of a gaseous product molecule, plus a molecule of solid product referred to as the nucleus of the reaction. This first step is repeated randomly, but a greater reaction is produced by the outwards growth of these nuclei, promoted by the strain put on the bonds around the product nucleus. This growth may be geometrically spherical, cylindrical or linear, often dependent upon the crystal structure. There is also interface growth, where a two-dimensional interface sweeps through a crystal of reactant from one end to the other. Another class of mechanism depends upon diffusion of one or more reactants or a gaseous product through the solid structure. This may also depend upon geometry and the crystal structure of the reacting mass. Various laws of diffusion have been applied, leading to a variety of possible mechanisms. Each of these mechanisms leads to a kinetic equation. As many as 29 equations have been proposed,[18] derived from these theories, but there is no consensus as to which ones are valid. A list of 18 commonly used equations is given in Table 4.

The notation column follows a system for coding the equations suggested by Sharp *et al.*[19] The name column gives the researchers who first suggested the equation or information about the mechanism involved in the derivation. These equations should be contrasted with the simple zero, 1st, 2nd and 3rd order equations from conventional kinetic studies. A fuller list may be found in reference 18. The first difficulty in studying solid-state kinetics is in deciding which of such a large number of equations is being obeyed. This is compounded because several equations may give very similar curves and thus appear to give equally good fits to some experimental data.

The starting point for examining kinetic data is to write the differential equation, as in conventional kinetics.

$$d\alpha/dt = k f(\alpha) \tag{5}$$

Table 4 *Some of the equations suggested for solid-state reactions*

No.	Integral equation $g(\alpha) = kt$	Differential equation $d\alpha/dt = kf(\alpha)$	Notation	Name
	Diffusion controlled			
1	$kt = \frac{1}{2}\alpha^2$	$d\alpha/dt = k\alpha^{-1}$	D1	One-dimensional
2	$kt = (1-\alpha)\ln(1-\alpha) + \alpha$	$d\alpha/dt = -k/\ln(1-\alpha)$	D2	Two-dimensional
3	$kt = \frac{3}{2}[1 - (1-\alpha)^{1/3}]^2$	$d\alpha/dt = k(1-\alpha)^{2/3}[1 - (1-\alpha)^{1/3}]^{-1}$	D3	Jander (three-dimensional)
4	$kt = \frac{3}{2}[1 - (2\alpha/3)] - (1-\alpha)^{2/3}$	$d\alpha/dt = k/[[(1-\alpha)^{1/3} - 1]$	D4	Ginstling–Brounshtein
	Sigmoidal α–time curve			
5	$kt = 2[-\ln(1-\alpha)]^{1/2}$	$d\alpha/dt = k(1-\alpha)[-\ln(1-\alpha)]^{1/2}$	A2	Avrami–Erofeev
6	$kt = 3[-\ln(1-\alpha)]^{1/3}$	$d\alpha/dt = k(1-\alpha)[-\ln(1-\alpha)]^{1/3}$	A3	Avrami–Erofeev
7	$kt = \frac{4}{3}[-\ln(1-\alpha)]^{3/4}$	$d\alpha/dt = k(1-\alpha)[-\ln(1-\alpha)]^{1/4}$	A4	Avrami–Erofeev
8	$kt = \ln[\alpha/(1-\alpha)]$	$d\alpha dt = k\alpha(1-\alpha)$	B1	Prout–Tompkins
	Order with respect to α			
9	$kt = \alpha$	$d\alpha/dt = kC$	F0	Zero order
10	$kt = -\ln(1-\alpha)$	$d\alpha/dt = k(1-\alpha)$	F1	First order
11	$kt = (1-\alpha)^{-1} - 1$	$d\alpha/dt = k(1-\alpha)^2$	F2	Second order
	Geometric models			
12	$kt = 2[1 - (1-\alpha)^{1/2}]$	$d\alpha/dt = k(1-\alpha)^{1/2}$	R2	Interface (contracting area)
13	$kt = 3[1 - (1-\alpha)^{1/3}]$	$d\alpha/dt = k(1-\alpha)^{2/3}$	R3	Interface (contracting volume)
14	$kt = \frac{3}{2}[1 - (1-\alpha)^{2/3}]$	$d\alpha/dt = k(1-\alpha)^{1/3}$	R4	Interface
	Power law			
16	$kt = 2\alpha^{1/2}$	$d\alpha/dt = k\alpha^{1/2}$	P1($n=2$)	Power law (half power)
17	$kt = 3\alpha^{1/3}$	$d\alpha/dt = k\alpha^{2/3}$	P2($n=3$)	Power law (third power)
18	$kt = 4\alpha^{1/4}$	$d\alpha/dt = k\alpha^{3/4}$	P3($n=4$)	Power law (quarter power)

In this α is the fraction reacted, t is time, k is the rate constant, and $f(\alpha)$ is the particular kinetic function, for example taken from those in Table 4.

The experiment could be carried out isothermally or non-isothermally. These methods will be considered separately.

Isothermal Kinetics

This technique is easy to conduct, as explained above, but suffers from the difficulty of deciding at what time the sample reached the reaction temperature.

One method of analysis would be to take the original data, which is a plot of α *versus* t, and to differentiate it to obtain $d\alpha/dt$. The values of $d\alpha/dt$ would then be plotted against $f(\alpha)$. The slope would be the value of k, the rate constant. There would actually have to be 18 plots, and a decision would have to be made as to which is the best straight line. As mentioned above, the differentiation process tends to promote "noise" and, with experimental scatter of points, makes this a difficult technique.

An alternative analysis derives from integrating the basic equation. If equation (5) is transposed and integrated it becomes:

$$\int d\alpha / f(\alpha) = \int k \, dt \qquad (6)$$

The left-hand side is the algebraic integral of $1/f(\alpha)$ and is represented by $g(\alpha)$. This term represents the 18 different equations in Table 4, one from each of the 18 $f(\alpha)$ functions. If k is a constant, which it will be at constant temperature, the right-hand side will become kt, so the integrated equation becomes:

$$g(\alpha) = k \, t \qquad (7)$$

The original data, without differentiation, may then be inserted into $g(\alpha)$ and plotted against time, t. Again the difficulty is that 18 plots must be made and a decision made about the best fit to a straight line. There is now less "noise", but the decision is still difficult.

An alternative is to use reduced time plots.[19] In this a computer was used to generate standard sets of data points for an imaginary reaction with some suitable value of rate constant k, arbitrarily chosen. This was repeated for all of the 18 model mechanisms, producing 18 tables of data. The time scales of these tables and the experimental data were completely different. However, time scales could be standardised by finding the time when a standard fraction of the reaction was complete. The obvious point is when half of the reaction is over or $\alpha = 0.5$. This time is referred to as $t_{0.5}$. The times in the tables were then all divided by this time to produce $t/t_{0.5}$, which is referred to as reduced time. This was repeated for each table of standard and experimental data so that they now all had the same time scale. The alpha scale in each case was the same, running from 0.0 to 1.0. Graphs could then be drawn of α against $t/t_{0.5}$. The plots were curves, not straight lines, but were spread out, and it was relatively easy to see which standard line the experimental line closely fitted. The lines were shown more widely spread out if the standard or matching point was taken at $\alpha = 0.9$ rather than at 0.5. The reduced time scale was then $t/t_{0.9}$. In practice, 18 theory lines and one experimental line on one plot are too much to distinguish from one another, so it is better to have five or six theory plots only on each graph, plus the experimental one. Once the equation number was decided upon, it was relatively easy to plot equation (7) to determine k; or even to calculate $g(\alpha)/t$ and average the values to get k.

The Arrhenius equation states that:

$$k = A\exp(-E/RT) \tag{8}$$

where R is the molar gas constant.

If the experiment is conducted at several constant temperatures, then an Arrhenius plot of $\ln(k)$ *versus* $1/T$ (where T is absolute temperature) will give activation energy E since the slope is $-E/R$, and the natural log of the pre-exponential factor A is the intercept.

Non-isothermal Kinetics

Since T is continuously rising in this type of experiment, the value of k is also rising. This would appear to allow the possibility of deriving E and A from one single experiment instead of a set of experiments as in the isothermal method above. To integrate equation (5) this time requires that k is replaced by $A\exp(-E/RT)$ giving:

$$\int d\alpha / f(\alpha) = \int A \exp^{-E/RT} dt \tag{9}$$

Since the heating rate is constant, the temperature at any time t is given by;

$$T = T_0 + \beta t \tag{10}$$

where β is the heating rate and T_0 is the starting temperature. Differentiating gives:

$$dT = \beta dt \quad \text{or} \quad dt = dT/\beta$$

Equation (9) then becomes:

$$\int d\alpha / f(\alpha) = \int \frac{A}{\beta} \exp^{-E/RT} dT \tag{11}$$

The left-hand side is again $g(\alpha)$ but unfortunately the right-hand side is a non-integrable function. Various simplifications have been tried, and equations that approximate to the values required over short ranges have been suggested.[20] These have had varying success. The reader should refer to the suggested texts for fuller discussions on the difficulties of this technique.

One worry about using a rising temperature technique is that the mechanism being followed may change as temperature rises, thus invalidating the whole technique.

Example Experiment

A simple example of a kinetic experiment is described here. This was conducted to find the mechanism and kinetic constants for the decomposition of calcium oxalate monohydrate by an isothermal experiment.

As shown in Figure 6(A), the decomposition of this compound takes place in three stages. It was the middle stage, the decomposition of the anhydrous material, which was to be studied. It was necessary to remove the water of crystallisation before measuring the kinetics of the second stage. Around 10 mg of the sample was weighed out onto the thermal balance, the furnace was raised into position and the temperature raised to, and held at 180°C, which was sufficient to drive off the water. The temperature was held until constant weight was achieved. The next move was to raise the temperature to a level to cause the oxalate to decompose to the carbonate. The furnace could not be lowered, in order to raise it to reaction temperature, because the sample picked up water rapidly from the atmosphere. With some care with the temperature control, it was possible to achieve a rapid rise without any overshoot of the required temperature. Some readings were inevitably taken after decomposition started, but before the constant temperature was reached. The point where the final temperature was achieved had to be found and the data edited to remove the earlier points. This was done in the analysing software on the computer. Modern control systems may appear to take care of the temperature control automatically, but it should not be assumed that this control is perfect and the temperature readings should be examined to see how quickly the temperature settles at the required value. The furnace temperature was held until constant mass was achieved for a second time. Logging of data was then stopped and the apparatus switched off.

Data points were taken every 5 s and consisted of time, thermocouple EMF and mass readings. The mass data for this one step of reaction were converted into fraction reacted, α. The EMF readings were averaged and converted into temperature in °C, taken as the constant temperature of the reaction. The computer program determined the point of standard reaction, in this case where 0.9 of the reaction was complete. The time for this event was divided into all the logged times to obtain the reduced time, $t/t_{0.9}$. A plot was then made of the α values against reduced time. On the same graph the standard data for possible mechanisms that might match the experimental result were also plotted. To allow clarity of the plots, in practice the standard data was divided into several separate plots. *One* plot only is shown in Figure 12(A).

The standard data is represented as 6 sets showing points only. The equations (9) to (14) are taken from Table 4. There are 40 points in each plot at intervals of α of 0.025. The continuous line is the experimental result. Equation (10), represented by upside down triangles, may be seen to be the best fit. Equation (11) shows points that are too high and the rest all lie too low. The other graphs are not shown, but all gave poor fits between points and line. Numerical deviation factors between points and line were also calculated, listed for each equation. It was obvious from these factors, as well as the graph, that Equation (10) from Table 4 is the correct one. Equation (10) is known as mechanism F1, in the $f(\alpha)$ form it is $d\alpha/dt = k(1 - \alpha)$ and in the $g(\alpha)$ form it is $g(\alpha) = -\ln(1 - \alpha)$. The program also divided $g(\alpha)$ by t and averaged the result and displayed this as the rate constant k.

The experiment was repeated at a number of different temperatures, not too far from the original one. The experiment must not be too slow or too fast to allow proper data logging. The results are shown in Table 5 for six temperatures.

The quantities $\ln(k)$ and $1/T$ (T is temperature in K) were also entered into an EXCEL spreadsheet. This was used to plot $\ln(k)$ *versus* $1/T$ as an Arrhenius plot as shown in Figure 12(B). Minus the slope was multiplied by the gas constant to obtain the activation energy, which had a value of 165.9 kJ mol^{-1}. The intercept gave the value of $\ln(A)$ as 22.316. This means that A was 4.9×10^9 s^{-1}.

A further reference for reading is the standard method found in the

A

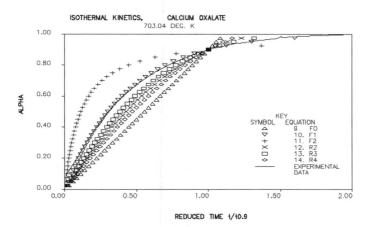

B

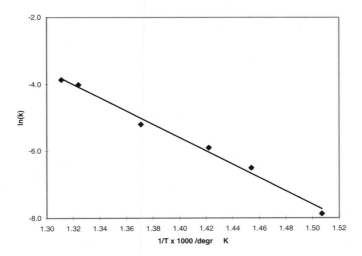

Figure 12 *Kinetics of decomposition of calcium oxalate.* (A) *Reduced time plot and* (B) *Arrhenius plot*

Table 5 *Rate constants at various temperatures for calcium oxalate decomposition*

Temperature/°C	Temperature /K	10^3 K$/T$	10^3k / s^{-1}	ln(k)
390.32	663.47	1.507	0.38	$-$ 7.87
414.82	687.97	1.454	1.49	$-$ 6.51
429.89	703.05	1.422	2.73	$-$ 5.90
455.98	729.13	1.371	5.51	$-$ 5.20
482.02	755.17	1.324	18.14	$-$ 4.01
489.84	762.99	1.311	21.02	$-$ 3.86

ASTM E 1641 (1999) test method for decomposition kinetics by thermogravimetry.

ANCILLARY TECHNIQUES

Very often thermogravimetry alone cannot give enough information about the reactions being studied. Other measurements often add to the knowledge gained. These ancillary techniques may be applied at the same time as the TG measurement is being applied and are then referred to as *simultaneous* and are discussed in Chapter 6. Alternatively other measurements may take place in separate experiments in separate apparatus. This is referred to as *combined* measurements. Simultaneous measurements include differential thermal analysis (DTA) and evolved gas analysis (EGA). These are explained in later chapters.

Residue Analysis

Combined measurements include examination of the residue from the crucible. This may include Fourier-transform infrared spectrometry, electron probe microanalysis, and scanning electron microscopy. A simple measurement is the determination of the X-ray diffraction pattern (XRD). This permits the identification of product materials by comparison with standard pattern tables. This method may fail if the product is a glass, as in the example above, and normally crystalline materials dissolve in the glass while it is molten. If the solid is affected by exposure to air in removing it to an X-ray camera (*e.g.* oxidation), then a high temperature X-ray diffraction may be used. The specific surface area of the powder residue may be found by gas adsorption techniques, usually using N_2 gas at liquid N_2 temperature. The area is expressed as m^2 g^{-1}. It has been found that, if there is a large difference in molar volume between reactant solid and product solid, the strains set up during decomposition tend to shatter the particle into smaller pieces, causing a large increase in area. This will also show up if particle size is measured. Changes in the method of heating can affect the increase in area found. For instance cobalt and nickel oxalate hydrates will dehydrate to porous or non-porous oxalate depending upon conditions. This will affect the subsequent decomposition. Following on from this it may be seen that displacement density of the solid may produce useful results. If a residue is heated to a high temperature, there is a tendency for sintering to occur, with subsequent loss of area. Thus it is useful not to go to too high a temperature but to stop just after decomposition.

Other measurements that have been applied to the solid include measurements of reflectance spectra during decomposition. Similarly, magnetic susceptibility and electrical conductivity have been used. Another technique is to place the sample on the heated stage of a microscope and to directly observe a reaction interface progressing across the sample.

Other Temperature Regimes

Temperature may not always be raised in a linear fashion. In the case of CRTA (Controlled Rate Thermal Analysis), the heating rate is varied in such a manner as to produce a constant rate of mass loss.[21,22] Alternatively a sinusoidal temperature rise is superimposed on the linear rise; this is known as Modulated TG[23] and allows the continuous calculation of activation energy and pre-exponential factor during a run. Sometimes a Temperature Jump (or stepwise isothermal)[24] is used, where temperature is held constant for a time, then jumped rapidly to a higher constant temperature (usually quite close in temperature). All of these procedures are supposed to help in the determination of kinetics of reaction. Another system accelerates the temperature rise when no mass loss is experienced, *i.e.* between reactions. The rate is slowed to a low value during mass loss. Some manufacturers call this High Resolution TG and an example follows.

High Resolution Thermogravimetry

To separate reactions which occur at temperatures close to one another, a high resolution is required. Using a very slow, constant heating rate with conventional TG, Figure 13 shows that the decompositions of sodium bicarbonate, $NaHCO_3$ and potassium bicarbonate, $KHCO_3$ which both occur in the temperature range 110 to 200°C are almost resolved if a heating rate of 1°C min^{-1} is used. Unfortunately, this means that the TG experiment takes nearly 3 h.

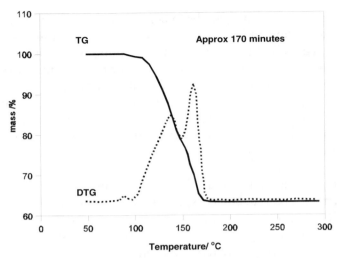

Figure 13 *Bicarbonate mixture. Conventional TG at* $1\,^{\circ}C\ min^{-1}$

Various approaches to improving resolution while keeping the duration of the experiment to a few minutes have been suggested.[25,26] One commercial implementation allows the use of four experimental regimes:

- Conventional constant heating rate.
- A stepwise isothermal mode where the sample is heated until a reaction is detected and then held constant until that reaction is complete.
- Constant reaction rate, where the heater is controlled to achieve a constant rate of mass change, for example 1% min^{-1}.
- Dynamic rate which varies the heating rate smoothly and continuously in response to the rate of sample decomposition, so that the resolution of the mass change is maximised. This mode allows rapid heating in regions where no transitions occur, but slows down the heating proportionally during a reaction. The effect of this is shown in Figure 14 where the bicarbonate reactions are better resolved in an experimental time of only 14 min.[27,28]

FURTHER READING

C. J. Keattch and D. Dollimore, *An Introduction to Thermogravimetry*, Heyden, London, 2nd edn., 1975.
J. G. Dunn and J. H. Sharp, "Thermogravimetry", in *Treatise on Analytical Chemistry*, ed. J. D. Winefordner, John Wiley & Sons, New York,

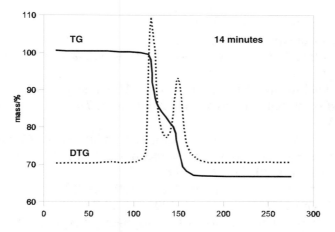

Figure 14 *Bicarbonate mixture. Hi-Res TG (dynamic rate)*

1993, Part 1, Vol. 13, pp. 127–266.
J. G. Dunn, "Thermogravimetry" in *Encyclopedia of Analytical Chemistry*, ed. R. A. Meyers, John Wiley & Sons, Chichester, 2000, Vol. 15, pp. 13 206–13 226.

REFERENCES

1. R. C. Mackenzie, "Nomenclature in Thermal analysis", in *Treatise on Analytical Chemistry*, ed. P. J. Elving and I. M. Kolthoff, John Wiley & Sons, New York, 1983, Part 1, Vol. 12, pp. 1–16.
2. J. G. Dunn and J. H. Sharp, "Thermogravimetry", in *Treatise on Analytical Chemistry*, ed. J. D. Winefordner, John Wiley & Sons, New York, 1993, Part 1, Vol. 13, pp. 127–266.
3. C. J. Keattch, *Talanta*, 1967, **14**, 77.
4. A. Savitzky and M. J. E. Golay, *Anal. Chem.*, 1964, **36**, 1627; J. Steiner, Y. Termonia and J. Deltour, *Anal. Chem.*, 1972, **44**, 1906; H. H. Madden, *Anal. Chem.*, 1978, **50**, 1383.
5. D. Dollimore, D. L. Griffiths and D. Nicholson, *J. Chem. Soc.*, 1963, 2617.
6. D. Dollimore and G. R. Heal, *Carbon*, 1967, **5**, 65.
7. S. J. Gregg and W. B. Jepson, *J. Inst. Metals*, 1958–59, **87**, 187. D. W. Aylemore, S. J. Gregg and W. B. Jepson, *J. Electrochem. Soc.*, 1960, **107**, 495.
8. ASTM E 1131 (1998), *Compositional Analysis by Thermogravimetry*, ASTM, West Consohocken, PA, USA.
9. C. M. Earnest and R. L. Fyans, *Perkin-Elmer Application Study* #32,

Perkin-Elmer, Beaconsfield.

10. M. R. Ottaway, *Fuel*, 1982, **61**, 713.
11. J. O. Hill, E. L. Charsley and M. R. Ottaway, *Thermochim. Acta*, 1984, **72**, 251.
12. C. F. Dickinson, PhD Thesis, Salford University, 2000.
13. J. Šesták, V. Šatava and W. W. Wendlandt, *Thermochim. Acta*, 1973, **7**, 333.
14. M. E. Brown, D. Dollimore and A. K. Galwey, in *Comprehensive Chemical Kinetics*, ed. C. H. Bamford and C. F. H. Tipper, Elsevier, Amsterdam, 1980, Vol. 22, 340 pp.
15. M. E. Brown, *J. Thermal. Anal. Cal.*, 1997, **49**, 17.
16. J. H. Flynn, *J. Thermal. Anal.*, 1995, **44**, 499.
17. J. H. Flynn, "The Historical Development of Applied Non-Isothermal Kinetics", in *Thermal Analysis*, ed. H. G. Wiedemann, Birkhauser, Basel, 1972, Vol. 2, 3.
18. C. F. Dickinson and G. R. Heal, *Thermochim. Acta*, 1999, **340–341**, 89.
19. J. H. Sharp, G. H. Brindley and B. N. Narahari Achar, *J. Am. Ceram. Soc.*, 1966, **49**, 379.
20. G. R. Heal, *Thermochim. Acta*, 1999, **340–341**, 69.
21. J. Rouquerol, "Controlled Transformation Rate Thermal Analysis: The Hidden Face of Thermal Analysis", *Thermochim. Acta*, 1989, **144**, 209.
22. M. Reading, "Controlled Rate Thermal Analysis and Beyond", in *Thermal Analysis – Techniques and Applications*, ed. E. L. Charsley and S. B. Warrington, Royal Society of Chemistry Special Publications, No. 117, Cambridge, 1992, pp. 126–155.
23. N. Buckmann, R. Blaine and G. Dallas, *Chem. Aust.*, 1998, March, 22–23.
24. O. Toft Sørenson, *Thermochim. Acta*, 1981, **50**, 163.
25. J. Rouquerol, *Bull. Soc. Chim.*, 1964, 31.
26. F. Paulik and J. Paulik, *Anal. Chim. Acta*, 1971, **56**, 328.
27. C. J. Lundgren, *Int. Lab.*, March, 1992, 26.
28. T A Instruments Hi-Res™ TGA Application Brief, T A Instruments, Leatherhead.

Chapter 3

Differential Thermal Analysis and Differential Scanning Calorimetry

P. G. Laye

Centre for Thermal Studies, University of Huddersfield, UK

INTRODUCTION

Differential thermal analysis (DTA) and differential scanning calorimetry (DSC) are the most widely used of all the thermal analysis techniques. The concept underlying the techniques is simple enough: to obtain information on thermal changes in a sample by heating or cooling it alongside an inert reference. Historically the techniques have their origin in the measurement of temperature. Figure 1 is a schematic representation of the main parts of an instrument. The sample and reference are contained in the DTA/DSC cell. Temperature sensors and the means of heating the sample and reference are incorporated in the cell. Other terms which have been used to describe this part of the instrument include "specimen holder assembly" and more recently "instrument test chamber". A single computer unit operates the various control functions, data capture and analysis. The term "differential" emphasises an important feature of the techniques: two identical measuring sensors are used, one for the sample and one for the reference, and the signal from the instrument depends on the difference between the response of the two sensors. In this way the signal represents the thermal change to be studied free from diverse thermal effects which influence both sensors equally. This has the considerable merit of allowing high sensitivities to be designed into instruments. The nature of the measuring sensors and the form of the instrument signal are discussed later in the chapter.

It is the link with thermal energy which is responsible for the wide

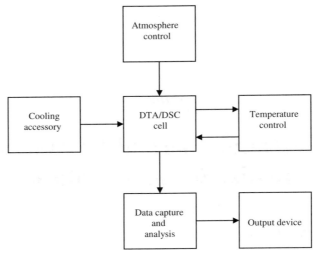

Figure 1   *Schematic representation of a DTA or DSC instrument*

ranging applicability of both DTA and DSC. Unlike thermogravimetry the techniques are not dependent on the sample undergoing a change in mass. DSC is the more recent technique and was developed for quantitative calorimetric measurements. DTA does not lend itself to such measurements and has progressively been replaced by DSC even for measurements in the range 750–1600°C, which at one time were the sole province of DTA. DTA still finds application in the measurement of characteristic temperatures and in the qualitative identification of materials. It remains the technique for measurements above 1600°C where the high temperatures impose considerable design constraints on equipment. The techniques are most readily applied to the study of solids and whilst their application to liquids is not uncommon more careful attention to experimental practice is required. The small size of samples, often only a few mg, and the rapidity with which experiments can be carried out have played an important part in establishing the popularity of the techniques.

The present chapter concentrates on the basic features of the techniques and the use of commercial equipment to make various measurements. Theoretical considerations have been limited to setting out the essential ideas. In contrast experimental procedures are discussed at some length. It is not possible to derive explicit working equations which permit the interpretation of experimental data by anything approaching first principles. However, calibration may be used to circumvent limitations in the theory and forms the basis of all quantitative measurements. In this context it is important to remember that no determination can be more precise than the calibration and this remains true even with the use of software packages which enable the results to be obtained by the press of the proverbial button (or key!).

DEFINITIONS AND NOMENCLATURE

The *practical* distinction between DTA and DSC is in the nature of the signal obtained from the equipment. In the case of a differential thermal analyser it is proportional to the temperature difference,

$$\Delta T = T_{\text{S}} - T_{\text{R}},$$

established between the sample and an inert reference when both are subjected to the same temperature program. The subscripts S and R indicate the sample and reference respectively. In the context of this chapter the signal from a differential scanning calorimeter will be regarded as proportional to the difference in thermal power between the sample and reference, $d\Delta q/dt$.

The classical temperature program is a linear temperature change with respect to time. Complex programs can be implemented by combining different heating or cooling rates with isothermal periods. An example is stepwise heating which may be used to detect the onset of melting under quasi-isothermal conditions. An important innovation has been to overlay the linear temperature change with a regular modulation. This technique has become known as modulated temperature DSC (MTDSC), or occasionally as temperature modulated DSC (TMDSC). In the original form of the technique introduced in 1993[1] the modulation was sinusoidal. Other temperature variations which have been introduced by manufacturers have included square wave and saw-tooth modulations. MTDSC has been shown to have a number of advantages over conventional DSC, including increased sensitivity and resolution and the ability, in some systems, to separate multiple thermal effects.

Most differential scanning calorimeters fall into one of two categories depending on their operating principle: power compensation or heat flux.

Power-compensation: This term is applied to the design which in its original form was introduced in 1964.[2] Figure 2(a) shows the main features of the DSC cell – the provision of separate temperature sensors and heaters for the sample and reference. In the event of a temperature difference arising between the sample and reference, differential thermal power is supplied to the heaters to eliminate the difference and to maintain the temperature at the program value. The differential thermal power is the source of the instrument signal.

Heat flux: In this case the instrument signal is derived from the tem-

**Separate sample and reference temperature
sensors and furnaces**

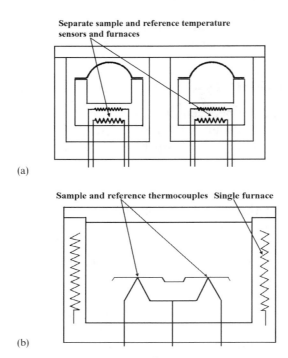

(a)

Sample and reference thermocouples Single furnace

(b)

Figure 2 (a) *Power-compensation differential scanning calorimeter.* (b) *Heat flux differential scanning calorimeter*

perature difference established when the sample and reference are heated in the same furnace. The temperature difference is measured by the temperature sensors – usually thermocouples arranged back-to-back. Figure 2(b) shows the arrangement of the thermocouples and the single furnace. The difference between heat flux DSC and DTA lies in the conversion of ΔT into differential power. The algorithm for this conversion is contained in the instrument software. The design of the DSC cell is critical if the algorithm is to be transferable from one experiment to another, independent of the sample. The heat flux approach is a development of older forms of "quantitative" DTA.

Both types of differential scanning calorimeters make use of a crucible to contain the sample. The reference is either an inert material in a crucible of the same type as that used for the sample or simply the empty crucible. Crucibles commonly measure 5–6 mm in diameter, which gives some idea of the overall dimensions of the DSC cell.

It is the provision of dynamic conditions in which the sample is subjected to a controlled heating or cooling program which sets DSC apart from other calorimetric techniques and is a key factor in its wide range of different applications. Most differential scanning calorimeters are of the heat flux type. Recently a differential scanning calorimeter has been designed which incorporates features of both power compensation and heat flux instruments.

The results from DTA and DSC experiments are displayed as a *thermal analysis curve* in which the instrument signal is plotted against temperature – usually the sample temperature – or time. Figure 3 shows some of the terminology relating to the results from DSC experiments. The description "heat flow" is frequently used for the instrument signal. Analysis of the thermal analysis curve is carried out using the instrument software. Of particular importance is the *extrapolated onset temperature* T_e which is defined as the temperature of intersection between the extrapolated initial base line and the tangent or line through the linear section of the leading edge of the peak. This temperature rather than the peak maximum temperature T_m is frequently used to characterise peaks because it is much less affected by the heating rate. The temperatures T_i and T_f are the initial and final temperatures of the peak which are sometimes more difficult to pin-point precisely. The terms "isothermal" and "dynamic" refer to the operating mode of the instrument. In the figure the peak represents an exothermic event (*exotherm*) and has been represented as a positive displacement. This is the usual convention for DTA and heat flux DSC. In the case of power-compensation DSC exotherms are negative displacements. Where confusion is likely to arise the direction of the exotherms/endotherms should be shown on the thermal analysis curve. In the present chapter the thermal analysis curves have been represented

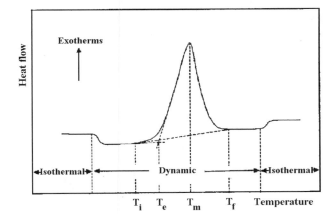

Figure 3 *Schematic representation of a thermal analysis curve*

using the "DTA convention" with the exception of Figure 7 where theory requires a positive displacement for the endotherm of fusion.

The alternative format for the thermal analysis curve is the plot of the instrument signal against time. The quantitative advantage of DSC over DTA lies in the relationship between the area enclosed by peaks measured in this format and the corresponding heat change: unlike DTA, with DSC the proportionality is independent of the heat capacity of the sample.

APPLICATIONS

The versatility of DTA and DSC can be seen in both the range of materials studied and the type of information obtained. From the standpoint of materials studied the techniques may be regarded as virtually universal in their applicability. Some idea of the range is given in Table 1. Although DSC is a quantitative technique it finds application alongside DTA as a qualitative tool whereby the thermal analysis curve is used as a fingerprint for identifying substances. The techniques may form part of a quality control procedure in which the presence or absence of a peak in the thermal analysis curve is all that is relevant. The identification of polymorphs in the context of pharmaceuticals is particularly relevant since different species may have quite different physiological actions. The investigation of potential reactivity between components of drugs as revealed by changes in the thermal analysis curve represents another significant application of these techniques. DTA and DSC have found valuable application in the study of phase diagrams both in pharmaceuticals and more widely in the general area of material science.

The greatest impact of the techniques has been seen in the study of polymeric materials with crystallinity and melting behaviour, glass transitions, curing processes and polymerisation representing the different types of thermal behaviour under investigation. A measure of the importance of this area of activity is reflected in the considerable number of publications and conference presentations it has generated.

Figure 4 illustrates the application of DSC to the study of polymers.

Table 1 *Materials studied by DTA and DSC*

Polymers, glasses and ceramics	Pharmaceuticals
Oils, fats and waxes	Biological materials
Clays and Minerals	Metals and alloys
Coal, lignite and wood	Natural products
Liquid crystals	Catalysts
Explosives, propellants and pyrotechnics	

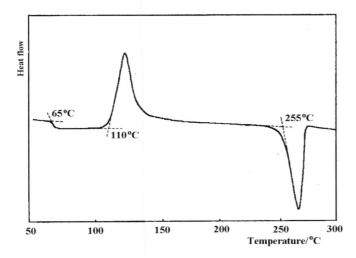

Figure 4 *Thermal analysis curve for poly(ethylene terephthalate)*

The curve is for poly(ethylene terephthalate) (PET) heated in N_2 at 10°C min^{-1}. The curve shows a displacement at 60–70°C arising from the glass transition and peaks at extrapolated onset temperatures of about 110°C and 255°C for crystallisation and melting respectively. In the figure the transitions are particularly well defined. However, both the physical properties and the composition and history of polymers affect glass transitions and crystallisation. The temperature of the glass transition is not fixed but is dependent on the rate of heating. The displacement often shows additional structure making the analysis much less straightforward than in the present example. At higher temperatures the thermal analysis curve would reveal the onset of degradation – thermal analysis offers a comparatively simple way of assessing the relative thermal stability of polymers.

The study of clays and minerals played an important role in the development of DTA as an investigative technique. In some instances DTA and DSC provide one of the few routes to the identification of minerals. An ingenious method,[3] elegant in its simplicity, whereby the identity of a component in a mixture can be confirmed is to add the suspected mineral to the *reference crucible* and repeat the thermal analysis experiment. The relevant peaks should show a diminution in size.

One of the most celebrated examples of the application of DTA and DSC to material science, which had a public impact some 30 years ago, is shown in Figure 5. It relates to the use of high alumina cement which in some circumstances led to the weakening and in extreme cases the

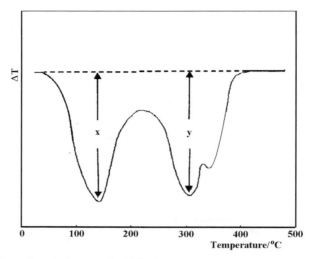

Figure 5   *Thermal analysis curve for high alumina cement*

collapse of concrete structures. The problem arose from the conversion of the initial product of setting $CaO \cdot Al_2O_3 \cdot 10H_2O$ into the hexahydrate and gibbsite (hydrated alumina). The issue became acute and DTA and DSC offered a rapid route to the determination of the extent of conversion. Many thousands of determinations were carried out to identify those cases where further structural investigations were needed. In the figure the endothermic peaks are for the dehydration of the decahydrate, gibbsite and hexahydrate in order of increasing temperature. The extent of conversion is defined as the amount of gibbsite/(amount of decahydrate + gibbsite). The "height" of the peaks is taken as a measure of the amounts so that the extent of conversion becomes $y/(x + y)$. In practice a calibration is carried out using a sample of known composition. The use of peak heights instead of the more correct use of areas to determine amounts is convenient where peaks overlap. DTA and DSC continue to find application in the investigation of the complex reactions in cement systems.

Figure 6 illustrates the application of DSC to the determination of the oxidative stability of oils.[4] Figure 6(a) shows the thermal analysis curve for the isothermal test in which the time to oxidation (t_{oxid}) is measured when the sample is maintained at a constant temperature in an atmosphere of O_2. An alternative test is dynamic where it is the temperature of oxidation which is measured. Specialised equipment is necessary for the isothermal test, which usually employs a pressure of about 3.5 MPa. An obvious advantage of both tests is that the performance of oils can be

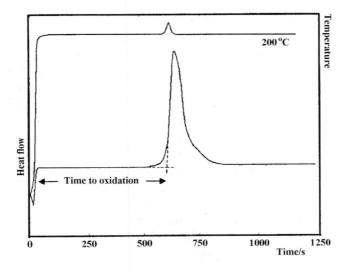

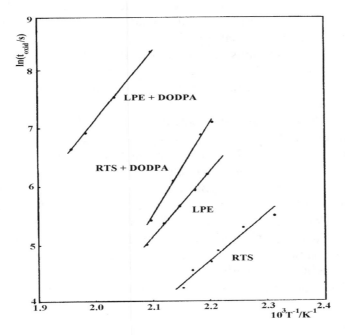

Figure 6   *Oxidative stability of oils.* (a) *Isothermal test experiment.* (b) *Determination of the activation energy of the induction reaction* (*abbreviations defined in the text*)

monitored without recourse to expensive and time-consuming engine tests. Furthermore it is possible to study the catalytic effects of metal surfaces on the oxidation. The reciprocal of time to oxidation has been taken as a measure of the rate of an induction reaction which has allowed the kinetics to be explored. Figure 6(b) shows the results for an ester base stock (LPE) and a synthetic base stock made from oligomerisation of dec-1-ene (RTS) and mixtures containing 1.5% by mass of the anti-oxidant dioctyldiphenylamine (DODPA). The activation energies range from 70 to 130 kJ mol^{-1}. Regardless of the precise significance of these values the true focus of interest lies in the variation from one base stock to another, with and without the addition of antioxidants.

Table 2 lists some of the quantitative measurements that can be undertaken by DSC. With modern equipment and software some of these have become largely a matter of routine. Even so some thought is necessary! Computer software has the propensity to produce "answers" to an impressive number of figures which may well be totally unrealistic. It is all too easy to forget that calorimetric measurements involve thermodynamic principles and that only by adhering to these can properly defined thermodynamic quantities be obtained. In spite of these reservations many determinations can be carried out with a minimum of difficulty. The American Society for Testing and Materials (ASTM) has described in considerable detail the use of DTA and DSC for a number of different measurements.

The use of DSC to investigate chemical kinetics deserves special mention. It has excited more interest and more controversy than perhaps any other area of application. It continues to generate an enormous output of literature. The basis for obtaining kinetic parameters is to identify the rate of reaction with the DSC signal and the extent of reaction with the fractional area of the peak plotted against time. It is possible to obtain the three variables, rate of reaction, extent of reaction and temperature by carrying out a series of isothermal experiments at different temperatures in much the same way as in classical kinetic investigations. The experimental procedure is not without its difficulty but the interpretation of the results is less contentious than with the alternative dynamic procedures.

Table 2 *Quantitative measurements by DSC*

Heat capacity
Enthalpies of transitions and transformations
Purity
Chemical kinetics
Vapour pressure
Thermal conductivity

However, it is these very procedures which exploit the unique capability of DSC. There has been something of a drive towards obtaining kinetic constants from a single dynamic experiment. Although the results obtained in this way may fulfil a useful function further tests are invariably needed to explore the possibility of limitations to their applicability. Advantages have been claimed for *sample controlled* kinetic experiments in which the experimental conditions are varied in order to maintain the rate of reaction constant. This has proved a popular method of temperature control in thermogravimetry although in principle it can be applied to DSC.

A standard test method for the determination of the kinetic constants for thermally unstable substances is one of the procedures published by ASTM E 698 (1999). The development of the test originally in 1979 highlighted the use of DTA and DSC for the investigation of potentially hazardous materials. This is an area of application which has continued to gain in importance with the heightened awareness of safety issues. Once again it is the need for small samples and the rapidity of the experiments which makes the techniques invaluable often as a preliminary to longer term and larger scale experiments such as adiabatic storage tests. Software is available which allows the assignment of kinetic parameters and the calculation of hazard potential. ASTM E 1231 (1996) describes standard practice for calculating the time-to-thermal runaway, critical half-thickness, critical temperature and adiabatic decomposition temperature rise.

THEORETICAL CONSIDERATIONS

Theoretical considerations can provide useful pointers to the interpretation of thermal analysis curves, can account for many of the empirical observations and offer guidance for good experimental practice. A theoretical approach requires spatial and temporal descriptions of the heat flux within the DTA/DSC cell from all forms of heat transfer across all interfaces. It is hardly surprising that explicit working equations cannot be derived. However, a great deal can be achieved using a simple approach which has the advantage of being easy to visualise. Such an approach was that adopted by Gray[5] where well established heat transfer equations were used to obtain expressions for DTA and power compensated DSC signals. Whilst Gray's analysis was concerned with both techniques our attention will be focused on DSC.

The aim was to derive an expression for the instrument signal in response to the evolution of heat from a sample as represented by dh/dt. The sample and its crucible were considered as one with a total heat capacity C_S. A similar assumption was made regarding the reference material and its crucible, which together had a total heat capacity C_R. It was assumed that there is a source of thermal energy at temperature T_p and a single thermal impedance R between the sample and the source of thermal energy and between the reference and the source of thermal energy. The heat flow between the thermal energy source and the sample was represented as dq/dt as measured by the instrument. The heating rate was represented by $dT_p/dt = \beta$ and assumed to be linear. Gray obtained the equation,

$$dh/dt = -dq/dt + (C_S - C_R)dT_p/dt - RC_S\, d^2q/dt^2. \tag{1}$$
$$\quad\quad\quad\quad\ \text{I}\quad\quad\quad\quad\quad\ \text{II}\quad\quad\quad\quad\quad \text{III}$$

Thus the heat evolution from the sample is given by the instrument signal measured from zero (term I), a heat capacity displacement (term II) and a third term which includes the product RC_S. This product has units of time so that term III represents a thermal lag. Included in the publication was a recipe for obtaining dh/dt from the experimental curve by making allowance for thermal lag. For inert samples $dh/dt = 0$ and the displacement (term II) provides a route to the determination of heat capacity.

The model serves to focus attention on the need to reduce thermal lag as much as possible. Although the contribution to thermal lag from the instrument is fixed by the nature of its design the practitioner has some control over the contribution from the sample and crucible. For example, the use of small samples and slow heating or cooling rates, good contact between the sample and crucible and between the crucible and the temperature sensor will all reduce thermal gradients.

Gray also discussed the shape of the leading edge of the peak for a melting transition,

$$dq/dt = (C_S - C_R)dT_p/dt + R^{-1}(dT_p/dt)t. \tag{2}$$
$$\quad\quad\quad\quad\quad\ \text{I}\quad\quad\quad\quad\quad\quad\ \text{II}$$

Thus the gradient depends on the product of R^{-1} and the heating rate (dT_p/dt) (term II) and provides a method which has been used to correct for thermal lag in assigning temperatures. Figure 7 shows the melting curve as described in Gray's theory.

Amongst the plethora of papers which have followed Gray's work that of Baxter[6] is interesting in that it presents a different perspective on DSC. Using a similar approach to that of Gray but involving two thermal impedance terms Baxter was able to relate the heat flux DSC signal to ΔT. Bearing in mind that differential power in power compensation DSC

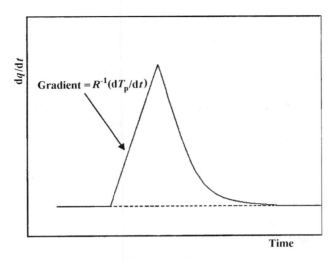

Figure 7   *Theoretical shape of a melting endotherm*

arises in response to an outer balance ΔT the distinction between the two types of DSC becomes blurred. For most purposes the results from the two approaches can be regarded as indistinguishable.

MTDSC represented something of a revolution in thermal analysis with an impact which has been compared to that of the original introduction of power compensation DSC. Numerous publications have appeared devoted to the complexity of the theory and the data manipulation techniques needed before useful information can be obtained. Fortunately all the "hard work" is done by the instrument software. Rather like conventional DSC, useful information can be obtained without recourse to the detailed theory.

The starting point adopted by Reading and co-workers[1] is a description of the heat flow into the sample which occurs as a result of the sinusoidal modulation of the temperature program,

$$dq/dt = C_s\beta + f(t,T). \qquad (3)$$
$$\;\;\;\;\;\;\;\;\;\;\;\;\;\;\;\; \text{I} \;\;\;\;\; \text{II}$$

Term I represents the heat capacity component of the signal. It is assumed that this component follows the periodically changing heating rate and is referred to as the reversing signal. The term $f(t,T)$ is any kinetically hindered thermal event and is regarded as non-reversing. Conventional DSC provides a measure of the total thermal power dq/dt whereas MTDSC allows the two components to be determined. The terms reversing and non-reversing relate to the conditions of the experiment. For example the sample size and the period of modulation will influence the ability of the sample to follow the temperature modulation and hence the apparent value of the heat capacity. It follows that the choice of experimental conditions is critical in MTDSC, far more so than in conventional DSC since it involves the selection of the period and amplitude of the modulation in addition to the underlying heating rate. To some extent these variables are interdependent. The aim is to achieve 4–6 cycles during the thermal event of interest The underlying heating rate can be set to zero in which case measurements are carried out under quasi-isothermal conditions. The software produces an output of the experimentally measured modulated heat flow and the modulated heating rate which can be used to judge whether the experiment has been carried out under satisfactory control. A plot of modulated heat flow against temperature should show a smooth modulation.

MTDSC has become very much the established technique in the study of polymeric materials where its advantages over conventional DSC can be exploited. Figure 8 illustrates the use of MTDSC for the separation of overlapping thermal events in a poly(ethylene terephthalate)–acrylonitrile butadiene styrene (PET–ABS) blend. The thermal analysis curve for PET alone was shown in Figure 4. The present figure shows the total heat flow for the blend and its separation into the reversing and non-reversing

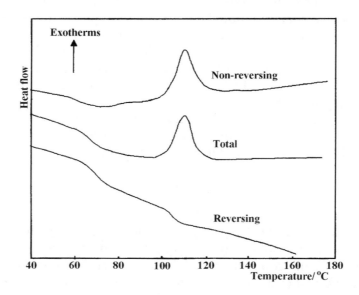

Figure 8   *Modulated DSC curve from a sample of PET–ABS blend. (The heat flow scales are different for the three curves)*

components. The glass transition temperature for PET is shown at about 65 °C and for ABS at 105 °C. Crystallisation of PET is shown as a peak in the non-reversing curve and is also at about 105 °C, which would mask the glass transition of ABS in conventional DSC.

INSTRUMENTATION

Specification

The aim here is simply to present an overview of the various features on offer. The range of instruments extends from differential scanning calorimeters in a suitcase for on-site use to spatially resolved micro-thermal analysis equipment for samples as minute as $2 \times 2\,\mu$m. Between these rather extreme examples there is a wide choice of commercial DTA and DSC equipment which allows samples to be studied at temperatures ranging from -150 °C to about 1600 °C. For higher temperature measurements (above 1600 °C) the equipment becomes increasingly more specialised. The detailed specification of equipment is often difficult (sometimes impossible!) to decipher – there appears to be no common practice between manufacturers. Information can best be obtained by raising questions directly with the manufacturers. Even so, hands-on experience is to be recommended when choosing equipment.

Temperature Sensors

In power compensated DSC the small size of the individual sample and reference holders makes for rapid response. The temperature sensors are platinum (Pt) resistive elements. The individual furnaces are made of Pt/Rh alloy. It is important that the thermal characteristics of the sample and reference assemblies be matched precisely. The maximum operating temperature is limited to about 750 °C. High temperature DSC measurements (750–1600 °C) are made by heat flux instruments using thermocouples of Pt and Pt/Rh alloys. The thermocouples often incorporate a plate to support the crucible. The use of precious metal thermocouples is at the expense of a small signal strength. Both chromel/alumel and chromel/constantan are used in heat flux DSC equipment for measurements at temperatures to about 750 °C. Multiple thermocouple assemblies offer the possibility of an increased sensitivity – recently a 20-junction Au/Au-Pd thermocouple assembly has been developed. Thermocouples of W and W/Re are used in DTA equipment for measurements above 1600 °C. The operating temperature is the predominant feature which determines the design and the materials used in the construction of the DTA/DSC cell. The performance of thermocouples can change with time due to chemical contamination and mechanical stress and as a result it is not possible to adopt a "once and for all" calibration.

Crucibles

Mention has already been made that samples can be as small as a few mg although such a size can raise considerable problems in obtaining a representative sample. The choice of crucibles depends on the construction of the DTA/DSC cell, the reactivity of the sample and the temperature range over which measurements are to be made. The most frequently used crucibles for low and moderate temperatures, $-150\,°C$ to about $600\,°C$, are made of aluminium and can be of a shallow or deep design. The temperature, $600\,°C$, is still well below the melting temperature of aluminium ($660\,°C$) but at higher temperatures there is the risk of irreversible (and very expensive!) damage to temperature sensors arising from alloying reactions. Lids are available which can be crimped into position using a specially designed press. In some experiments, such as measuring boiling temperatures, a fine hole is pierced in the lid. Crimped crucibles can withstand an internal pressure of about $0.3\,MPa$. Crucibles made of platinum are used for high temperature measurements but it should be remembered that platinum is not entirely resistant to chemical attack at high temperatures. Crucibles made of silver, gold, quartz, alumina and graphite are all available commercially. Also available are stainless steel crucibles with lids which screw down on a gasket. These will retain a pressure of $10\,MPa$ but are relatively massive ($1\,g$) and considerably increase the effective response time of the equipment. More specialised are glass capillary tubes which fit snugly inside a small metal block that takes the place of a conventional crucible. Liquid samples can be distilled into the capillary tubes or introduced *via* a syringe. Introducing a solid can be much more time-consuming! The tubes are sealed using a very fine flame whilst cooling the sample.

Temperature and Atmosphere Control

Computer software is responsible for the entire operation of equipment – at one time the emphasis seemed to be more on presentation of results. Heating rates from 0.1 to $500\,°C\,min^{-1}$ can be selected depending on the particular equipment: the high rates are seldom used in measurements but do allow temperature regions of interest to be reached quickly. Even so Mathot and co-workers[7] have pointed out that there is a need for much higher heating and cooling rates in order to investigate polymers under conditions similar to those used in processing. High cooling rates are also important since these increase the through-put of work – often an important commercial consideration. All equipment is designed to allow close control of the atmosphere in the DTA/DSC cell. Most experiments are done with a flowing atmosphere using flow rates from 10 to 100 cm^3 min^{-1}. Atmospheres may be reactive or inert but in the case of reactive atmospheres the possibility of reaction with materials used in the construction of the DTA/DSC cell needs to be taken into account. Some equipment has the in-built facility to switch from one gas to another during the course of an experiment.

Cooling Systems and Accessories

A variety of optional cooling systems is available from manufacturers to meet different needs. They are essential if controlled heating or cooling experiments are to be carried out below room temperature. More modest cooling may be used to enhance control at room temperature or just above. In MTDSC, cooling may be needed to ensure that the selected temperature modulations are attainable. Cooling systems are also used to fast-cool equipment following high temperature experiments. In this context a simple electric fan may be all that is necessary or "top-cooling" by means of a cooled metal block which fits into the DTA or DSC cell. A number of manufacturers have marketed robotic systems for automatic loading of samples. Commercial equipment is also available for high pressure DSC (HPDSC) in which the entire DSC cell can be pressurised to about 7 MPa and for photoDSC in which samples can be illuminated with UV radiation in order to initiate reactions. Most thermal analysers can be linked to analytical equipment such as mass spectrometers and some allow both DSC or DTA and thermogravimetry to be carried out simultaneously on the same sample.

Calvet-type Equipment

An alternative approach to the design of differential scanning calorimeters is to place the sample and reference inside a relatively massive calorimetric block which acts as a heat sink. The main features of the apparatus are shown in Figure 9. The instrument signal is obtained by measuring the heat flux between the sample and calorimetric block, and the reference and calorimetric block using thermopiles. The temperature signal is derived from a sensor in the block. The design based on that of Calvet has been used in calorimeters operating at a single temperature but is also used in DSC instruments where the temperature of the block

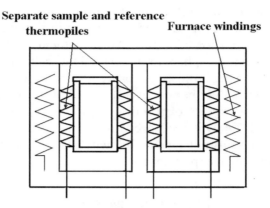

Figure 9 *Calvet-type differential scanning calorimeter*

may be raised or lowered. This type of DSC has the advantage of being able to accommodate very much larger samples and allows for more versatile experiments. The increased sample size and stable temperature environment increases the effective sensitivity of the apparatus but the downside is a very much longer response time and very slow heating and cooling rates. Mackenzie[8] has provided a comprehensive discussion of the relationship between the design of equipment and the nature of the instrument signal.

EXPERIMENTAL CONSIDERATIONS

Variables

Wendlandt[9] identified some 16 variables which influence the results from DTA and DSC experiments. Whilst many are attributable to the design of the equipment or to the inherent properties of the sample there remains a core of variables where the practitioner is able to exert some control. Sample preparation and containment, heating rate and atmosphere all come within this core and even small refinements in technique can often enhance the quality of the results. Two somewhat extreme examples illustrate the need for careful control of experimental technique. The first is the well-known example of the decomposition of an oxalate to carbonate and CO. This is an endothermic process but the thermal analysis curve will show an exotherm due to combustion of CO if there is a trace of air remaining in the apparatus. The second example illustrates how the use of a crucible with an ill-fitting lid may give rise to ambiguous results. Thus, in the study of emulsion explosives the vaporisation of the sample can lead to an extreme distortion in the shape of the exotherm.

Do's and Do not's

Crucibles should be clean and free from all traces of contaminants from their manufacture. Pre-treating the crucible by heating over the temperature range of the experiment beforehand is a useful ploy for ensuring the absence of unwanted signals during the experiment. The nature of a crucible surface may be another variable that needs to be considered. There are situations where the crucible may act as a catalyst as in the measurement of oxidation temperatures of oils.

Remember the importance of ensuring good contact between the sample and crucible and between the crucible and the thermocouple or other measuring sensor. It may be possible to flatten the base of the crucible by pressing it against a flat metal block. Whenever practicable, samples should be pressed into the crucible, perhaps by means of a lid or even a second crucible.

Contact between a sample and crucible in fusion experiments can be improved by pre-melting the sample to give a uniform coating over the base of the crucible. This has the effect of enhancing the shape of the subsequent endotherm with the proviso that no reaction has occurred. However, even here care is needed: the opposite effect is observed when powdered gold is melted to give globules which have poor thermal contact with the crucible.

A liquid may be introduced into a crucible using a syringe. Care is needed not to wet the outside of the crucible, particularly when crimping a lid into place. A tall sided crucible may be selected where there is a tendency for a liquid to spread over the surface or if reaction products tend to creep.

When a crucible is to be used more than once in a series of measurements it must be replaced each time in precisely the same position. One such example is in the measurement of heat capacity. Thermocouple assemblies are often designed to facilitate the repositioning of the crucible.

A cautionary "don't" is to pre-treat the sample without prior knowledge of the relevance of its history. Grinding a sample to a fine powder to give good contact with the crucible may introduce spurious thermal effects. The form of crystals, their size and shape, may be important factors in the kinetic behaviour of a sample.

When selecting an experimental procedure bear in mind the previous comments regarding aluminium crucibles and avoid their use above

600 °C. Be aware that alloying reactions can also occur when using platinum crucibles. Take into account the possibility of corrosive attack on thermocouples by some reaction products. A well-known example is HCl, which is evolved in the degradation of chlorinated polymers and readily attacks some thermocouple combinations even at modest temperatures. Poor reproducibility of the base line is often an indication of a contaminated cell. Don't attempt to clean DTA and DSC cells by scouring the surfaces. Gentle cleaning by brushing and heating in air should be all that is necessary. If alloying has taken place it is probably too late to retrieve the situation.

Be aware of the need to examine thermal analysis curves before subjecting the results to detailed analysis. Spurious signals are all too frequent and can arise for a number of reasons – samples bubbling, a spasmodic escape of gas from the crucible and even the movement or distortion of the crucible itself.

The choice of temperature program often involves a compromise. The desire for a large signal by use of large samples or high heating (or cooling) rates may lead to an unacceptable thermal lag. However, thermal lag is less of a problem now with the high sensitivity of modern equipment which allows the use of small samples. A heating rate of $10 \, °C$ min^{-1} is commonly used – at least in a preliminary investigation. Experiments should be started some $30 \, °C$ below the temperature of interest so that a quasi-steady-state can be established before making measurements.

Figure 10 shows the effect of heating rate on the fusion peak of indium displayed against temperature (a) and against time (b). The curves illustrate the efficacy of extrapolated onset temperature compared with the peak temperature as already discussed.

The nature and flow rate of the atmosphere in the DTA/DSC cell can have a significant effect on the thermal analysis curve. With samples in an open crucible the atmosphere can be reactive but is more commonly inert – N_2 or Ar being used to blanket the sample and sweep out evolved gaseous products. The atmosphere also plays a significant role in heat exchange and control of the flow rate is important even in those experiments where the sample is in a sealed crucible. The reduction of thermal lag remains as a priority in MTDSC. In spite of the expense, the high conductivity of He makes it the preferred choice of atmosphere in MTDSC. Fine control of the flow rate becomes even more important with the use of He. Changes in the nature of the atmosphere, whether by design or during the course of a reaction, may affect the thermal conductivity sufficiently to alter the peak shapes. A dry atmosphere must be used for measurements below room temperature otherwise condensation will

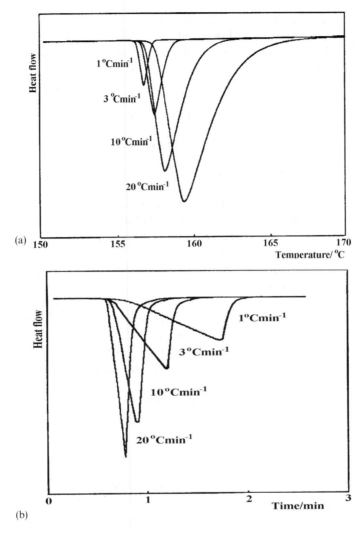

Figure 10 *The influence of heating rate on the thermal analysis curves for the fusion of indium:* (a) *plotted against temperature,* (b) *plotted against time*

occur in the sample holder, giving rise to spurious signals. With some equipment it may be advantageous to mount the DTA/DSC cell in a dry box.

Reference

The choice of a reference presents few difficulties. In the past, calcined alumina (heated to 1500 °C to remove adsorbed moisture) was commonly used and still remains the favoured choice when samples are large. For small samples the tendency is to use no reference material at all, merely an empty crucible – as mentioned earlier.

CALIBRATION

Preliminaries

So far the comments concerning experimental aspects have been in general terms – this is the part of the chapter where they become more specific. Although calibration is primarily concerned with temperature and energy (or enthalpy) it is also the stage at which the practitioner gets to know the equipment. It is a good idea to start with preliminary experiments as a precursor to the detailed calibration.

It is false economy to "short-change" calibration. The point has been made that it is not a once and for all procedure. However, following a comprehensive calibration it is usually only necessary to carry out periodic checks. It seems to be the case that equipment functions better if its use is limited to a restricted range of temperatures rather than being used for all temperatures – but this has financial implications! It is important to keep a full record of the calibration procedures, periodic checks and all the results.

Preliminary experiments should aim to answer basic questions about the behaviour of equipment:

(1) How reproducible is the instrument signal when a sample crucible is removed and subsequently replaced in the apparatus?
(2) How sensitive is the signal to the precise placing of the crucible in the apparatus?
(3) How sensitive is the signal to the atmospheric flow rate?
(4) How does temperature affect these experiments?

To some extent at least the answers may be an indication of the prowess of the practitioner.

The resolution of the equipment can be examined by recording the thermal analysis curve for a mixture of substances which have close transition temperatures – the greater the resolution the sharper the distinction between the two peaks. Quartz and potassium sulfate with solid-state transition temperatures of 573 °C and 583 °C respectively are often used for this purpose.

It is important to check the signal range for which the instrument response is linear. This may be carried out by recording the fusion peak of a suitable calibrant – indium is often used – and determining the mass range for which the area of the peak is directly proportional to the sample mass. With modern equipment the response is likely to be linear over an extensive range of operation.

Temperature Calibration

The object is to assign the correct temperature to the temperature indicated by the instrument. The method depends on recording the thermal analysis curve for substances which exhibit a suitable transition temperature. Substances of well established transition temperatures are available which cover much of the temperature range of DSC instruments. It is important that any substance used as a calibrant should be available in pure form, be stable and if possible have a negligible vapour pressure. Many of the substances are metals and some have temperatures of melting which are fixed points on the International Practical Temperature Scale. Organic calibrants should always be used in sealed crucibles to avoid any loss of sample through vaporisation. This will ensure a sharp peak and at the same time will avoid damage to the equipment from vapour. Table 3 lists some of the substances for which the temperatures are either fixed points or are well established.

A standard procedure for the temperature calibration of differential thermal analysers and differential scanning calorimeters has been published as ASTM E 967 (1999). In the two point method two calibrants are chosen to bracket the temperature range of interest. It is assumed that the correct temperature T is related to the experimental temperature T_{exp} by the relationship,

$$T = T_{\text{exp}} S + I. \tag{4}$$

S and I are defined by the relationships,

Table 3 *Calibration standards*

	$T_{\text{trans}}/°C$	$\Delta_{\text{trans}}H^{\theta}/\text{J g}^{-1}$
Cyclopentane, s → s	− 135.1	4.91
Cyclopentane, s → l	− 93.4	8.63
Gallium, s → l	29.8	79.9
Benzoic acid, s → l	123	148
Indium, s → l	156.6	28.6
Tin, s → l	231.9	60.4
Aluminium, s → l	660.3	398

$$S = (T_1 - T_2)/(T_{exp1} - T_{exp2}), \tag{5}$$

$$I = [(T_{exp1} \times T_2) - (T_{exp2} \times T_1)]/(T_{exp1} - T_{exp2}), \tag{6}$$

where the subscripts 1 and 2 refer to the two calibrants. S should be close to unity.

In principle the method is simple enough – small quantities of the two calibrants in turn are allowed to equilibrate 30 °C below the temperature of the transition as indicated by a constant instrument signal. The calibrants are then heated through the transition and the extrapolated onset temperatures obtained from the endotherms. These temperatures correspond to T_{exp1} and T_{exp2}. The calibration will be affected by heating rate and possibly the nature of the atmosphere and its flow rate. The experimental conditions for the calibration should be chosen to match those for subsequent measurements. Calibration over more than two points may be carried out and the relationship between T and T_{exp} determined statistically. The extent to which the instrument output can be corrected by the software will depend on the detailed design of the computer system.

The process of temperature calibration and measurement has been considered in considerable detail in connection with a program of work by GEFTA (German Thermal Analysis Society). The authors raise the fundamental issue that whereas the temperatures of the fixed points are defined for the substances in phase equilibrium the experimental results are measured under dynamic conditions. The authors recommend a procedure based on extrapolation of results to zero heating rate. The details are contained in a series of publications.[10,11] The authors[12] have also considered the problem of temperature calibration under conditions of decreasing temperatures.

An approach to obtaining the melting temperature under quasi-isothermal conditions is by stepwise heating (Figure 11). As the temperature is increased the peaks initially reveal only a heat capacity displacement. As the melting temperature is approached this displacement is combined with some pre-melting and finally at the melting temperature the fusion peak itself is obtained. Thereafter only a heat capacity displacement is observed. The procedure can be used to identify the experimental melting temperature to a fraction of a degree.

Figure 12 shows the variation of the solid–solid transition temperature of K_2CrO_4 with heating rate as part of a study into the feasibility of using K_2CrO_4 as a temperature standard.[13] The transition temperature of K_2CrO_4 was determined within a closely defined experimental protocol

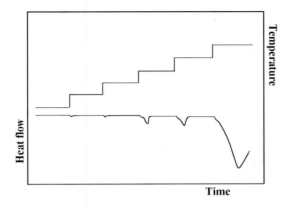

Figure 11 *Schematic representation of stepwise heating to obtain melting temperatures under quasi-isothermal conditions*

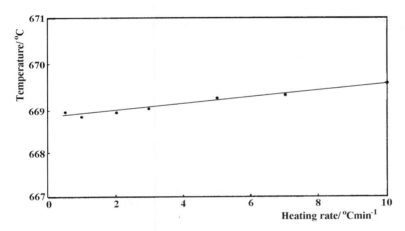

Figure 12 *Variation of the extrapolated onset temperature with heating rate for the solid–solid transition peak of K_2CrO_4*

in three laboratories on five different thermal analysers (both differential thermal analysers and differential scanning calorimeters). The value $669.1 \pm 0.2 \,°C$ was obtained relative to a single point calibration with aluminium for a heating rate of $3\,°C$ min^{-1} and an atmosphere of nitrogen.

Energy Calibration

The object here is to relate the instrument signal to thermal power. The use of electrical heating is not possible for the majority of differential scanning calorimeters with the result that calibration is dependent on the availability of chemical calibrants. A number of the temperature standards have well established enthalpies of transition, permitting both temperature and energy calibration to be performed together (Table 3). In practice the transitions employed are almost invariably fusions. The process of energy calibration has been reviewed by Sarge *et al.*[14] as part of the GEFTA programme of work on calibration.

The starting point is the relationship between the experimental thermal power $(d\Delta q/dt)_{exp}$ and the correct value $d\Delta q/dt$,

$$d\Delta q/dt = \kappa(d\Delta q/dt)_{exp}, \tag{7}$$

where κ is the calibration constant. The use of equation (7) implies that the instrument signal has already been converted into experimental thermal power and κ is dimensionless. Alternatively the instrument signal may be read simply as a voltage, in which case κ will have the units $J\ s^{-1}\ V^{-1}$.

The integral of the instrument signal with respect to time over a fusion peak is the total heat change,

$$\int (d\Delta q/dt)\ dt = Q = \kappa \int (d\Delta q/dt)_{exp}dt, \tag{8}$$

where Q is the heat change. It is assumed that κ does not vary with time (and hence temperature) over the temperature range of the peak – taking for granted that the peak is sharp. Thus κ can be determined from the heat change $Q = m\Delta_{fus}H$ where m is the mass of the sample and $\Delta_{fus}H$ is the specific enthalpy of the fusion.

The area of the fusion peak is obtained using the available software package but care is needed to ensure that the integration extends over the entire peak. The sample should be weighed using a balance capable of weighing to at least 0.01 mg. Results obtained in experiments where a weight loss occurs should be regarded as suspect. The inclusion of the determined calibration constant into the software once again depends on the design of the computer system. The calibration constant should be determined over the temperature range of interest in subsequent measurements.

The initial determination of the linearity of the instrument response will be important in determining the extent of sample mass or thermal power for which the calibration holds true. Even so it is sensible to investigate whether the calibration constant shows any dependence on heating rate, the nature and flow rate of the atmosphere, the type of crucible and the thermal characteristics of the sample – metal *versus* non-metal.

It is also possible to use the area enclosed by a peak which is obtained when a sample of known heat capacity is heated through an accurately known temperature range. Sapphire (α-alumina) finds almost universal application in this approach. It has a well established heat capacity over the working temperature range of differential scanning calorimeters and has the added advantage of being commercially available in the form of discs to fit snugly in crucibles. Alternatively the instrument signal can be calibrated directly – thermal power $d\Delta q/dt$ rather than through total heat Q. This alternative will be referred to later in the context of heat capacity measurements. The use of heat capacity measurements to extend the more conventional calibration based on enthalpies of fusion has been described by ASTM [E 968 (1999)]. Figure 13 shows the linear response of a differential scanning calorimeter by means of recording the instrument signal (in mV) for a number of discs of sapphire.

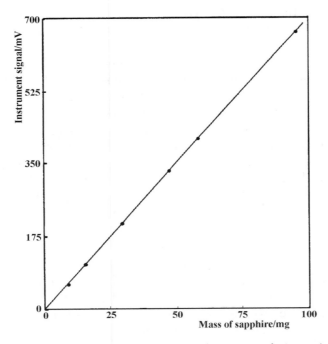

Figure 13 *Test for the linear response of a differential scanning calorimeter (see text for details)*

MEASUREMENTS

Reference has already been made to the range of applications of DTA and DSC. This section is concerned with a closer look at calorimetric measurements which link thermal power to heat capacity, $d\Delta q/dt = (C_S - C_R)\beta$, and its integral, $\int(d\Delta q/dt)dt$, to energy or enthalpy. These linkages together with temperature form the basis of quantitative DSC. Computer systems for the control of equipment, data capture and subsequent analysis have combined to give increased versatility and results of far greater precision than was possible previously with chart recorders.

Measurement of Heat Capacity

This represents one of the early successes of conventional DSC and more recently of MTDSC. Heat capacity is a key thermodynamic quantity because of its intrinsic importance and its relationship to other quantities – enthalpy, entropy and Gibbs energy. Its measurement continues to be an important application of DSC where results can be obtained having an uncertainty of only 1–2%, often with a minimum of difficulty. At the other end of the scale with rigorous attention to detail and with suitable samples results can be obtained with uncertainties approaching those of adiabatic calorimetry.[15]

The classical method of measuring heat capacities using DSC involves three experiments. In each experiment the isothermal base-line is established and the calorimeter then programmed over a temperature range before establishing the final isothermal base-line. Experiments can be carried out using an increasing or decreasing temperature. The same crucible is used for the three experiments and each experiment is carried out over the same temperature range and at the same heating or cooling rate. The DSC cell contains (1) the empty crucible, (2) the crucible + calibrant and (3) the crucible + sample. The reference, an empty crucible, is left undisturbed throughout the sequence of experiments.

Figure 14 shows the thermal analysis curves for the three experiments. The difference between the isothermal signals at the beginning and end of the experiments arises from the temperature dependence of the heat transfer coefficients. The isothermal signals from the three experiments have been adjusted to be coincident. Modern equipment with its more secure base-line enables the temperature range of the experiment to be 100°C or more. The selection of the sample mass is something of a compromise – a large sample will increase the magnitude of the signal but will give a greater uncertainty in the temperature due to increase in

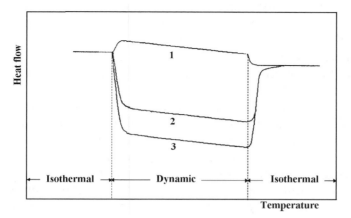

Figure 14 *Determination of heat capacity using the classical DSC method (see text for details)*

thermal lag. A sample mass of 10–20 mg has been recommended with a heating (or cooling) rate of 10–20 °C min^{-1}. As indicated in the previous section, sapphire is almost invariably used as the calibrant. To reiterate the point already made, the crucible should be in precisely the same position in the DSC cell for each experiment. A lid should be used for the crucible to cover the calibrant and sample and the DSC cell should be closed by a lid. The signal for each experiment at temperature T can be represented by the relationships,

1 $$(d\Delta q/dt)_1 = (C - C_R)\,\beta = \kappa\,(d\Delta q/dt)_{exp1} \tag{9}$$

2 $$(d\Delta q/dt)_2 = (C + C_C - C_R)\,\beta = \kappa\,(d\Delta q/dt)_{exp2} \tag{10}$$

3 $$(d\Delta q/dt)_3 = (C + C_S - C_R)\,\beta = \kappa\,(d\Delta q/dt)_{exp3}, \tag{11}$$

where C is the heat capacity of the empty crucible, C_C, C_S and C_R the heat capacity of the calibrant, sample and reference respectively.

The difference between the signals for experiments 1 and 2 serves to calibrate the thermal power,

$$\kappa = C_C\beta\,/[(d\Delta q/dt)_{exp2} - (d\Delta q/dt)_{exp1}]. \tag{12}$$

The heat capacity of the sample is obtained from the formula,

$$C_S = C_C[(d\Delta q/dt)_{exp3} - (d\Delta q/dt)_{exp1}]/[(d\Delta q/dt)_{exp2} - (d\Delta q/dt)_{exp1}]. \tag{13}$$

It is seen that the calibration constant disappears, which assumes that it is constant over the experimental conditions. The calculation is carried out using dedicated software. In some circumstances the crucible used for the sample may have to be different from that used for the calibrant. This means that a correction will be required to take into account the difference between the heat capacity of the two crucibles – readily calculated with sufficient accuracy. Measurements can be made at a series of temperatures but are meaningful only within the quasi-steady-state region of the experiment. The specific heat capacity of sapphire has been listed by ASTM in connection with the standard test method E 1269 (1999) for determining specific heat capacity by differential scanning calorimetry.

A number of difficulties can arise. The first is that it is rarely possible to adjust the isothermal signals at the beginning of the experiment and find that they are coincident at the end. Richardson[16] has summarised his series of papers dealing with this mismatch which may limit the temperature range of the experiments. The second difficulty is uncertainty in the assignment of temperature arising from thermal lag. Again Richardson[16] has indicated a method of assessing thermal lag from the area of the tail following the change from dynamic to isothermal mode of operation at the end of the experiment. It corresponds to an energy $C_s \delta T$ whereby the temperature correction can be calculated.

The measurement of heat capacity changes through glass transitions often presents difficulties when using the classical technique. The increase is small – often less than $1 \text{ J K}^{-1} \text{ g}^{-1}$ – but the importance of the glass transition is paramount marking as it does significant changes in the properties of polymeric substances. A number of alternative approaches have been devised to improve the resolution of glass transitions. These include the variety of modulated heating programmes. It was in the measurement of glass transitions that MTDSC made such an early impact. The approach differs from that of the classical route in that it depends on the comparison between thermal power recorded at different heating rates. In MTDSC the heating rate is continually changing as a result of the temperature modulation and the heat capacity is obtained by dividing the modulated heat flow by the modulated heating rate. The detailed calculation depends on a rather complicated manipulation of the raw data but again all this is in the software! The reversing component of the total heat flow is then obtained by multiplying the heat capacity by the average heating rate. A feature of the new techniques is that measurements can be carried out under quasi-isothermal conditions – impossible to achieve with the conventional technique. In this way it is possible to monitor the time dependence of the heat capacity during reactions.

Measurement of Energy

The area of peaks is determined by software with the input of the times corresponding to the start and finish of the peak. The software may provide different options for constructing the base-line. The calculation is straightforward if the peak is reasonably sharp and the heat capacity of the product does not differ greatly from that of the initial sample. A straight line from start to finish of the peak is constructed to represent the base-line. Some difficulty may occur in identifying the precise starting point of a peak – a problem often encountered when determining the crystallinity of a polymer from the crystallisation peak. The real complexity arises when there is a marked change in heat capacity and the base-line no longer even approximates to a straight line. The construction of the base-line must reflect the changing heat capacity of the sample which in turn will depend on the proportion of the initial sample (reactant) and the product. Recipes have been published[17,18] which enable the base-line to be calculated from an analysis of the shape of the peak and a knowledge of the heat capacity of the reactant and product over the temperature range of the peak. The heat capacity of reactant can be obtained by extrapolating the base-line recorded before the peak. A similar construction can be applied to the product or it may be possible to recycle it over the temperature range of the peak.

The energy of the isothermal event may be obtained conveniently using the construction shown in Figure 15 in which heat capacity, $C = (dq/dt)/\beta$, is plotted against temperature for an exothermic reaction. Area ABEA represents the energy change for the fractional conversion of reactant into product at temperature T. The area ABCDEA corresponds to the energy of complete conversion of reactant into product at temperature T. The variation of this area over the temperature range of the peak is the temperature dependence of the energy. The construction also provides a secure route to kinetic data. The total area corresponds to the transformation $\alpha = 0$ to $\alpha = 1$ where α is the fractional extent of reaction. The fractional area ABEA/ABCDEA is the extent of reaction α at temperature T. The values of α obtained in this way may be used to construct the curve AFD which represents the contribution to the instrument signal of the instantaneous heat capacity of the composition defined by α. The rate of reaction $d\alpha/dt$ at temperature T is proportional to the distance FE. Treatment of the entire curve yields $d\alpha/dt$, α and T – the triplet which forms the starting point for a conventional approach to obtaining a description of the kinetics. The derivation of kinetic data in this way assumes that the instrument is perfect, with no thermal lags.

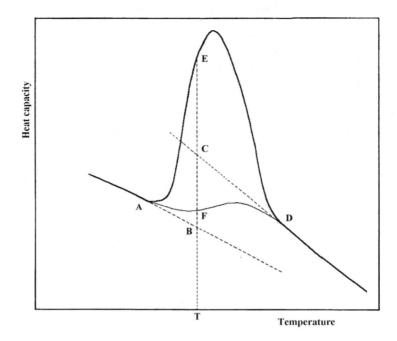

Figure 15   *Calculation of the isothermal energy change and the rate of reaction*

Measurement of Purity

The use of DSC for the measurement of purity depends on an analysis of the shape of the fusion peak of the substance. It has found fairly wide application but is limited to relatively pure substances (> 97 mol%). The results become less reliable for substances with greater levels of impurity. The method is non-specific – it is not a method for identifying the specific nature of impurities. It mirrors the classical approach which depends on the depression of the freezing temperature by the presence of an impurity. Figure 16 shows the thermal analysis curves for the melting of benzoic acid containing increasing amounts of an impurity. The curves show an increase in the width of the peaks – the temperature range over which melting takes place. The analysis depends on using fractional areas of the peak to calculate the extent of melting.

Ideal solution theory, which forms the basis of the method, involves a number of assumptions. In particular it applies only to weak solutions which in the context of the determination means nearly pure substances. Furthermore the impurity should not dissolve in the solid phase. The

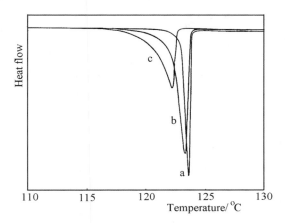

Figure 16   *Melting endotherms for benzoic acid*: (a) 99.9, (b) 99 and (c) 97 mol% purity

relationship between temperature (in kelvin) and the fraction of the sample melted f is given by

$$T = T_o - x(R T_o^2/\Delta_{fus}H)\, 1/f. \qquad (14)$$

T_o is the temperature of melting of the pure substance regarded as the solvent. $\Delta_{fus}H$ is the enthalpy of fusion and x is the mole fraction of the impurity. Clearly the results will not be meaningful if the sample decomposes at or near the melting temperature.

The experimental procedure is usually straightforward. The sample (1–3 mg) is enclosed in a sealed crucible and the fusion peak recorded using a slow heating rate, $1\,°C\ min^{-1}$ or less. Figure 17(a) shows a thermal analysis curve for a purity determination. $\Delta_{fus}H$ is calculated from the total area enclosed by the fusion peak (ACD) plotted against time. The fraction of the sample melted at the temperature T is obtained by dividing the partial area (ABE) by the total area of the peak. Some allowance is made for thermal lag in assigning the temperature using the construction shown in the figure. The gradient of the line BT is equal to that of the leading edge of the fusion peak of the calibrant, usually indium, recorded at the same heating rate. It is recommended that the calculation should extend over 10–50% of the total area of the peak.

The plot of temperature against $1/f$ should be a straight line but in practice a curve is obtained [ABC in Figure 17(b)]. Much controversy has surrounded the reason for this curvature with one explanation being that a small amount of initial melting is undetected. In any event a small

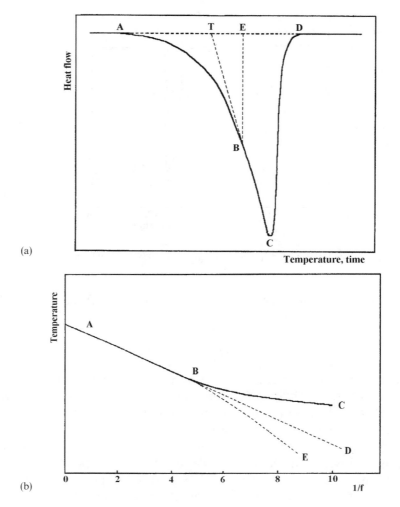

Figure 17   *Determination of purity:* (a) *fusion endotherm* (b) *plot of temperature against 1/f (see text for details)*

correction is added to all the areas to obtain a straight line, ABD. The correction is usually about 10% of the total area. The curve ABE is an example of over-correcting the initial curve. The mole fraction impurity is then calculated from the value of the gradient.

The method has proved popular, not least because of its rapidity and the simplicity of the analysis, and has been adopted by ASTM as a "standard test method" [E 928 (1996)]. Nowadays the analysis is included in the software package. A note of caution however: it may not be obvious whether the "ideal solution" theory represents the melting behaviour of a particular sample. As might be expected a number of efforts have been made to refine the method and extend its range of applicability. One such method is based on stepwise heating which has already been mentioned Although it detects melting within finer temperature limits it is much more time-consuming.

Kinetics

The application of thermal analysis to the study of kinetics involves so many ramifications that few would dispute that it could fill a book of its own. In principle at least the determination of kinetic parameters requires an investigation of the rate of reaction over all values of extent of reaction and temperature. Only in this way can potential changes in the reaction mechanism be identified. It cannot be accomplished by a single experiment but if the experimental conditions are sufficiently extensive the results are capable of providing useful information. Even so, extrapolation of rate data outside the conditions of the experiment needs to be undertaken with care. Results may be used to obtain predictive curves which relate extent of conversion, time and temperature. The isothermal law has been linked to a variety of mechanistic models but the ultimate determination of mechanisms depends on the input of results from a variety of techniques.

The experimental study of kinetics, as mentioned earlier, has as its basis the identification of the rate of reaction with instrument signal and the extent of reaction with fractional area of the peak. Analysis of experimental data often makes use of "named" approaches which exploit the advantages of dynamic experiments to achieve results without recourse to protracted experimental effort. Two popular but very different methods are those of Borchardt and Daniels[19] and Ozawa[20] and both appear in ASTM methods. Both are supported by commercial software. The concern here is with the Borchardt and Daniels method [ASTM E 2041 (1999)] which had a considerable impact on kinetic evaluation both by DTA and subsequently DSC. The analysis was devised originally for DTA experiments in which the thermocouples were in large volumes of stirred liquids. This is in direct contrast to the current application of the method to DSC studies of solids. A number of assumptions were made which were met more readily in stirred liquid systems than with solids. As a result there are a number of caveats associated with its application to solids. In particular the analysis assumed the absence of temperature gradients and thermal lag. From the standpoint of DSC this means small samples and slow heating rates – not the ideal experimental scenario to obtain a large signal! A sample mass of a few mg and heating rates no greater than $10\,^{\circ}\mathrm{C\ min^{-1}}$ should prove to be effective.

The Borchardt and Daniels approach is usually associated with an analysis based on equation (15):

$$\ln(d\alpha/dt) = \ln A - E_a/RT + \ln(1 - \alpha), \qquad (15)$$

which assumes first order kinetics. A and E_a are the pre-exponential factor and the Arrhenius activation energy respectively and the temperature T is in kelvin. As before, α is the fractional extent of reaction. The original assumption of first order kinetics was one of a number made to simplify the expression originally derived by Borchardt and Daniels for DTA. The rate of reaction $d\alpha/dt$ at a given temperature is obtained from the thermal power, $d\alpha/dt = (dq/dt)/\Delta H$ where ΔH may be obtained from the total area enclosed by the thermal analysis peak. The value of α is obtained from the fractional area at the corresponding temperature. A plot of $\ln[(d\alpha/dt)/(1 - \alpha)]$ against $1/T$ should be a straight line which confirms the order of reaction to be unity and leads to values for the activation energy and pre-exponential factor.

Once again the advantage of the method is its speed and simplicity. However, it is applicable only to first order reactions which lead to a smooth well shaped thermal analysis peak. Nevertheless be wary of using software to smooth thermal analysis curves – detail may be lost which might be critical in deciding whether the method is applicable. Although allowance should be made for the effect of thermal lag on the shape of the thermal analysis curve in practice it is seldom carried out. Instead the experimental procedure aims to minimise temperature gradients. ASTM recommend that the maximum heat evolution should be restricted to less than 8 mW. Furthermore the base-line is usually a straight line drawn from the start to the finish of the peak. The use of the approach has been extended by assuming a more general isothermal law $f(\alpha)$, whence equation (15) becomes

$$\ln(d\alpha/dt) = \ln A - E_a/RT + \ln f(\alpha). \qquad (16)$$

It is impossible to summarise the determination of kinetics on the basis of such a short section devoted to one approach. What is patently obvious is that the determination of kinetics by DSC is a veritable minefield: different methods often seem to lead to different results despite the best efforts on the part of practitioners.

Final Thoughts on Measurement

One Last Word of Warning: Be aware that different software or even up-grades of the same software can be a source of unexpected difficulty.

A Word of Encouragement: It is important to remember that DSC remains one of the easiest techniques to use. It is possible to perform numerous types of experiments with the minimum of difficulty. It is important not to get bogged down by complications that can occur – let the complications overtake you as and when they may – don't start with them!

FURTHER READING

Accounts of DTA and DSC can be found in the recently published Volume 1 of *Handbook of Thermal Analysis and Calorimetry*[21] and in *Differential Scanning Calorimetry, An Introduction for Practitioners*.[22] Both texts provide a valuable source of references to original literature. An older text is the book by Wendlandt[23] which like the Handbook is not restricted to DTA and DSC but nevertheless contains an enormous amount of information. Other texts are available which deal with specific areas of application: food,[24] materials,[25] petroleum analysis,[26] pharmaceuticals[27] and polymers.[28,29]

Several references have been made in this chapter to the importance of MTDSC. The increasing number of publications and conference presentations concerned with the use of MTDSC is a measure of its importance. Refs. 30 and 31 are to recent scientific meetings devoted specifically to MTDSC. Reference has also been made to sample controlled thermal analysis. This technique was developed independently by Paulik and Paulik[32] and by Rouquerol[33] and has been reviewed by Reading.[34] Kinetics has been touched on only briefly. Recent publications[35–38] have summarised the results of a kinetic project by the International Confederation for Thermal Analysis and Calorimetry (ICTAC) in which sets of kinetic data were distributed to participants for analysis using whichever methods they chose. The first of these publications provides background information. The isothermal kinetics of solid-state reactions have been the subject of a recent publication by Galwey and Brown.[39] Lists of calibrants, their melting temperatures and enthalpies of fusion are given in refs. 21 and 22. The issue of correcting for thermal lag is discussed at length in *Differential Scanning Calorimetry, An Introduction for Practitioners*.[22] Finally it is worth remembering that manufacturers are an excellent source of references to the literature.

And for the future, reference 40 is to a special issue of *Thermochimica Acta* entitled "Towards the New Century".

REFERENCES

1. M. Reading, D. Elliott and V. L. Hill, *J. Thermal Anal.*, 1993, **40**, 949.
2. E. S. Watson, M. J. O'Neill, J. Justin and N. Brenner, *Anal. Chem.*, 1964, **36**, 1233.
3. P. D. Garn, *Thermoanalytical Methods of Investigation*, Academic Press, New York, 1965, p. 606.
4. D. J. Rose, Analysis of Antioxidant Behaviour in Lubricating Oils, PhD Thesis, School of Chemistry, University of Leeds, 1991.
5. A. P. Gray in *Analytical Calorimetry*, ed. R. S. Porter and J. F. Johnson, Plenum Press, New York, 1968, p. 209.
6. R. A. Baxter, in *Thermal Analysis*, ed. R. F. S. Schwenker and P. D. Garn, Academic Press, New York, 1969, p. 65.
7. T. F. J. Pijpers, V. B. F. Mathot, B. Goderis and E. van der Vegte, *Proceedings of 28th* NATAS Oct 4–6, 2000, Orlando, Florida, USA.
8. R. C. Mackenzie, *Anal. Proc.*, 1980, 217.
9. W. Wm. Wendlandt, *Thermal Analysis*, Wiley-Interscience, New York, 3rd edn., 1985, 227.
10. G. W. H. Höhne, H. K. Cammenga, W. Eysel, E. Gmelin and W. Hemminger, *Thermochim. Acta*, 1990, **160**, 1.
11. H. K. Cammenga, W. Eysel, E. Gmelin, W. Hemminger, G. W. H. Höhne and S. M. Sarge, *Thermochim. Acta*, 1993, **219**, 333.
12. S. M. Sarge, G. W. H. Höhne, H. K. Cammenga, W. Eysel and E. Gmelin, *Thermochim. Acta*, 2000, **361**, 1.
13. E. L. Charsley, P. G. Laye and M. Richardson, *Thermochim. Acta*, 1993, **216**, 331.
14. S. M. Sarge, E. Gmelin, G. W. H. Höhne, H. K. Cammenga, W. Hemminger and W. Eysel, *Thermochim. Acta*, 1994, **247**, 129.
15. J. E. Callahan, K. M. McDermott, R. D. Weir and E. F. Westrum, *J. Chem. Thermodyn.*, 1992, **24**, 233.
16. M. J. Richardson, in *Compendium of Thermophysical Property Measurement Methods*, ed. K. D. Maglic, A. Cezairliyan and V. E. Peletsky, Plenum, 1992, Vol. 2, p. 519.
17. W. P. Brennan, B. Miller and J. C. Whitwell, *Ind. Eng. Chem. (Fund.)*, 1969, **8**, 314.
18. H. M. Heuvel and K. C. J. B. Lind, *Anal. Chem.*, 1970, **42**, 1044.
19. H. J. Borchardt and F Daniels, *J. Am. Chem. Soc.*, 1957, **79**, 41.
20. T. Ozawa, *J. Thermal Anal.*, 1970, **2**, 301.
21. M. E. Brown (ed.), *Handbook of Thermal Analysis and Calorimetry*,

Principles and Practice, Elsevier, Amsterdam, 1998, Vol. 1.

22. G. W. H. Höhne, W. Hemminger and H.-J. Flammersheim, *Differential Scanning Calorimetry, An Introduction for Practitioners*, Springer, Berlin, 1996.

23. W. Wm. Wendlandt, *Thermal Analysis*, Wiley-Interscience, New York, 3rd edn., 1985.

24. V. R. Harwalker and C. Y. Ma (ed.), *Thermal Analysis of Foods*, Elsevier, Amsterdam, 1990.

25. W. Smykatz-Kloss and S. St. J. Warne (ed.), *Thermal Analysis in the Geosciences*, Springer-Verlag, Berlin, 1991.

26. H. Kopsch, *Thermal Methods in Petroleum Analysis*, VCH, Weinheim, 1995.

27. J. L. Ford and P. Timmins, *Pharmaceutical Thermal Analysis: Techniques and Applications*, Ellis Horwood, Chichester, 1989.

28. V. B. F. Mathot (ed.), *Calorimetry and Thermal Analysis of Polymers*, Hanser, Munchen, 1994.

29. E. A. Turi (ed.), *Thermal Characterisation of Polymeric Materials*, Academic Press, New York, 2nd edn., 1997.

30. C. Schick and G. W. H. Höhne (ed.), Special Issue: Temperature Modulated Calorimetry, *Thermochim. Acta*, 1997, **304/305**, 1–378.

31. C. Schick and G. W. H. Höhne (ed.), Special Issue: Investigation of Phase Transitions by Temperature-Modulated Calorimetry, *Thermochim. Acta*, 1999, **330**, 3–200.

32. J. Paulik and F. Paulik, *Anal. Chim. Acta*, 1971, **56**, 328.

33. J. Rouquerol, *Bull. Soc. Chim. Fr.*, 1964, 31.

34. M. Reading, in *Thermal Analysis: Techniques and Applications*, ed. E. L. Charsley and S. B. Warrington, The Royal Society of Chemistry, Cambridge, 1992, p. 126.

35. M. E. Brown, M. Maciejewski, S. Vyazovkin, R. Nomen, J. Sempere, A. Burnham, J. Opfermann, R. Strey, H. L. Anderson, A. Kemmler, R. Keuleers, J. Janssens, H. O. Desseyn, Chao-Rui Li, Tong B Tang, B. Roduit, J. Malek and T. Mitsuhashi, *Thermochim. Acta*, 2000, **355**, 125.

36. M. Maciejewski, *Thermochim. Acta*, 2000, **355**, 145.

37. S. Vyazovkin, *Thermochim. Acta*, 2000, **355**, 155.

38. B. Roduit, *Thermochim. Acta*, 2000, **355**, 171.

39. A. K. Galwey and M. E. Brown, *Thermochim. Acta*, 1995, **269/270**, 1.

40. W. Hemminger (ed.) Special Issue: Towards the New Century, *Thermochim. Acta*, 2000, **355**, 1–253.

Chapter 4

Thermomechanical, Dynamic Mechanical and Dielectric Methods

D. M. Price

IPTME, Loughborough University, UK

INTRODUCTION AND PRINCIPLES

The dimensional and mechanical stability of materials is of paramount importance to their use in the everyday world where they may encounter a wide variation in temperature through design or by accident. Many polymers are processed at elevated temperatures so as to enable them to flow and be more amenable to fabrication. Food items are cooked, pasteurised or otherwise heated or frozen. Ceramics are fired so as to consolidate their final structure. The relationship between a material's dimensional and mechanical properties and its temperature is studied by the techniques described within this chapter and, due to common concepts, the effect of heat on the electrical properties of materials is also considered.

Thermomechanical Analysis and Thermodilatometry

Thermomechanical Analysis (TMA) can be defined as the measurement of a specimen's dimensions (length or volume) as a function of temperature whilst it is subjected to a constant mechanical stress. In this way thermal expansion coefficients can be determined and changes in this property with temperature (and/or time) monitored. Many materials will deform under the applied stress at a particular temperature which is often connected with the material melting or undergoing a glass–rubber transition. Alternatively, the specimen may possess residual stresses which have

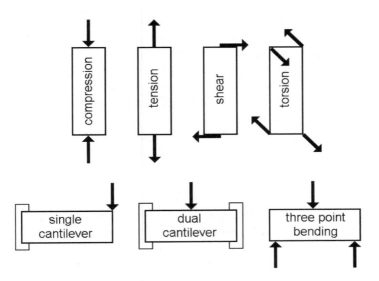

Figure 1 *Common mechanical deformation modes: compression, tension, shear, torsion, bending (single cantilever, dual cantilever, three point bending)*

been "frozen-in" during preparation. On heating, dimensional changes will occur as a consequence of the relaxation of these stresses.

Stress (σ) is defined as the ratio of the mechanical force applied (F) divided by the area over which it acts (A):

$$\sigma = F/A \qquad (1)$$

The stress is usually applied in compression or tension, but may also be applied in shear, torsion, or some other bending mode as shown in Figure 1. The units of stress are N m^{-2} or Pa.

If the applied stress is negligible then the technique becomes that of thermodilatometry. This technique is used to determine the coefficient of thermal expansion of the specimen from the relationship:

$$\alpha l_0 = dl/dT \qquad (2)$$

where α is the coefficient of thermal expansion (ppm °C^{-1} or μm m^{-1} °C^{-1}), l_0 is the original sample length (m) and dl/dT is the rate of change of sample length with temperature (μm °C^{-1}).

Dynamic Mechanical Analysis

Dynamic Mechanical Analysis (DMA) is concerned with the measurement of the mechanical properties (mechanical modulus or stiffness and damping) of a specimen as a function of temperature. DMA is a sensitive probe of molecular mobility within materials and is most commonly used to measure the glass transition temperature and other transitions in macromolecules, or to follow changes in mechanical properties brought about by chemical reactions.

For this type of measurement the specimen is subjected to an oscillating stress, usually following a sinusoidal waveform:

$$\sigma(t) = \alpha_{max}\sin \omega t \tag{3}$$

where $\sigma(t)$ is the stress at time t, σ_{max} is the maximum stress and ω is the angular frequency of oscillation. Note that $\omega = 2\pi f$ where f is the frequency in Hertz.

The applied stress produces a corresponding deformation or strain (ε) defined by:

$$\varepsilon = \text{(change in dimension)/(original dimension)} = \Delta l/l_o \tag{4}$$

The strain is measured according to how the stress is applied (*e.g.* compression, tension, bending, shear *etc.*). Strain is dimensionless, but often expressed as a %.

For an elastic material, Hooke's law applies and the strain is proportional to the applied stress according to the relationship:

$$E = d\sigma/d\varepsilon \tag{5}$$

Where E is the elastic, or Young's, modulus with units of N m^{-2} or Pa. Such measurements are normally carried out in tension or bending, when the sample is a soft material or a liquid then measurements are normally carried out in shear mode, thus a shear modulus (G) is measured. The two moduli are related to one another by:

$$G = E/(2 + 2v) \tag{6}$$

where v is known as Poisson's ratio of the material. This normally lies between 0 and 0.5 for most materials and represents a measure of the distortion which occurs (*i.e.* the reduction in breadth accompanying an increase in length) during testing.

If the material is viscous, Newton's law holds. The specimen possesses a resistance to deformation or viscosity, η, proportional to the rate of application of strain, *i.e.*:

$$\eta = d\sigma/(d\varepsilon/dt) \qquad (7)$$

The units of viscosity are Pa s.

A coil spring is an example of a perfectly elastic material in which all of the energy of deformation is stored and can be recovered by releasing the stress. Conversely, a perfectly viscous material is exemplified by a dash-pot, which resists extension with a force proportional to the strain rate but affords no restoring force once extended, all of the deformation energy being dissipated as heat during the loading process. In reality, most materials exhibit behaviour intermediate between springs and dash-pots – viscoelasticity.

If, as in the case of DMA, a sinusoidal oscillating stress is applied to a specimen, a corresponding oscillating strain will be produced. Unless the material is perfectly elastic, the measured strain will lag behind the applied stress by a phase difference (δ) shown in Figure 2. The ratio of peak stress to peak strain gives the complex modulus (E^*) which consists of an in-phase component or storage modulus (E') and a 90° out-of-phase (quadrature) component or loss modulus (E'').

The storage modulus, being in phase with the applied stress, represents the elastic component of the material's behaviour, whereas the loss

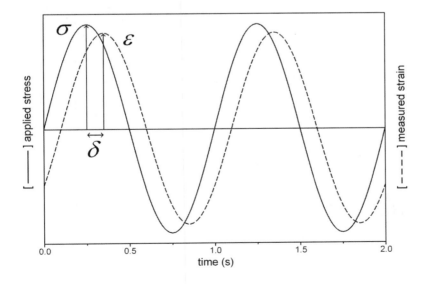

Figure 2 *Relationship between stress (σ) and strain (ε) during a dynamic mechanical test*

modulus, deriving from the condition at which $d\varepsilon/dt$ is a maximum, corresponds to the viscous nature of the material. The ratio between the loss and storage moduli (E''/E') gives the useful quantity known as the mechanical damping factor (tan δ) which is a measure of the amount of deformational energy that is dissipated as heat during each cycle. The relationship between these quantities can be illustrated by means of an Argand diagram, commonly used to visualise complex numbers, which shows that the complex modulus is a vector quantity characterised by magnitude (E^*) and angle (δ) as shown in Figure 3. E' and E'' represent the real and imaginary components of this vector thus:

$$E^* = E' + iE'' = \sqrt{(E'^2 + E''^2)} \tag{8}$$

So that:

$$E' = E^* \cos \delta \tag{9}$$

and

$$E'' = E^* \sin \delta \tag{10}$$

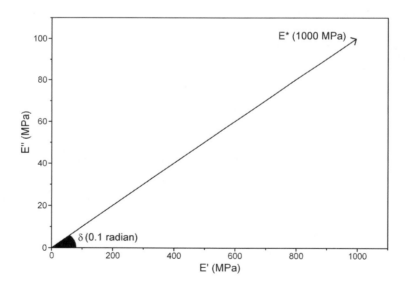

Figure 3 *Argand diagram to illustrate the relationship between complex modulus* (E*) *and its components*

Dielectric Techniques

In a manner analogous to TMA and DMA, a specimen can be subjected to a constant or oscillating electric field rather than a mechanical stress during measurements. Dipoles in the material will attempt to orient with the electric field, while ions, often present as impurities, will move toward the electrode of opposite polarity. The resulting current flow is similar in nature to the deformation brought about by mechanical tests and represents a measure of the freedom of charge carriers to respond to the applied field. The specimen is usually presented as a thin film between two metal electrodes so as to form a parallel plate capacitor. Two types of test can be performed.

Thermally Stimulated Current Analysis (TSCA)

In this technique the sample is subjected to a constant electric field and the current which flows through the sample is measured as a function of temperature. Often, the sample is heated to a high temperature under the applied field and then quenched to a low temperature. This process aligns dipoles within the specimen in much the same way that drawing a material under a mechanical stress would bring about orientation of molecules in the sample. The polarisation field is then switched off, and the sample is re-heated whilst the current flow resulting from the relaxation of the induced dipoles to the unordered state is monitored.

Dielectric Thermal Analysis (DETA)

In this technique the sample is subjected to an oscillating sinusoidal electric field. The applied voltage produces a polarisation within the sample and causes a small current to flow which leads the electric field by a phase difference (δ) (Figure 4). Two fundamental electrical characteristics, conductance and capacitance, are determined from measurements of the amplitude of the voltage (V), current (I) and δ. These are used to determine the admittance of the sample (Y) given by:

$$Y = I/V \tag{11}$$

Y is a vector quantity, like E^* discussed earlier, and is characterised by its magnitude $|Y|$ and direction δ.

The capacitance (C) is the ability to store electrical charge and is given by:

$$C = |Y| (\sin \delta)/\omega \tag{12}$$

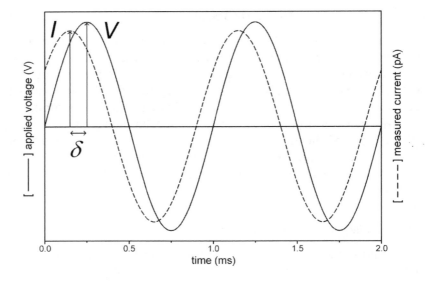

Figure 4 *Relationship between voltage and current in a capacitor (cf. Figure 2)*

The conductance (G_c) is the ability to transfer electric charge and is given by:

$$G = |Y| \cos \delta \qquad (13)$$

More usually data is presented in terms of the relative permittivity (ε') and dielectric loss factor (ε'') – these are related to capacitance and conductance by:

$$\varepsilon' = C/(\varepsilon_0.A/D) \qquad (14)$$

and

$$\varepsilon'' = G_c/(\omega.\varepsilon_0.A/D) \qquad (15)$$

where ε_0 is the permittivity of free space (8.86×10^{-12} F m^{-1}) and A/D, in m, is the ratio of electrode area (A) to plate separation or sample thickness, D, for a parallel plate capacitor. More generally, A/D is a geometric factor which is found by determining the properties of the measuring cell in the absence of a sample. Both ε' and ε'' are dimensionless quantities.

The ratio $\varepsilon''/\varepsilon'$ is the amount of energy dissipated per cycle divided by the amount of energy stored per cycle and known as the dielectric loss tangent or dissipation factor ($\tan \delta$).

INSTRUMENTATION

Thermomechanical Analysis

A schematic diagram of a typical instrument is shown in Figure 5. The sample is placed in a temperature controlled environment with a thermocouple or other temperature sensing device, such as a platinum resistance thermometer, placed in close proximity. The facility to circulate a cryogenic coolant such as cold nitrogen gas from a Dewar vessel of liquid nitrogen is useful for subambient measurements. The atmosphere around the sample is usually controlled by purging the oven with air or nitrogen from a cylinder. Because of the much larger thermal mass of the sample and oven compared to a differential scanning calorimeter or a thermobalance, the heating and cooling rates employed are usually much slower for TMA. A rate of $5\,^\circ\text{C}\,\text{min}^{-1}$ is usually the maximum recommended value for good temperature equilibration across the specimen. Even this rate can be a problem for some specimens where appreciable temperature gradients can exist between the middle and ends of the sample, particularly around the test fixtures – which can represent a significant heat sink.

For compression measurements (as illustrated) a flat-ended probe is rested on the top surface of the sample and a static force is applied by means of a weight or (more commonly in the case of modern instrumentation) an electromagnetic motor similar in principle to the coil of a

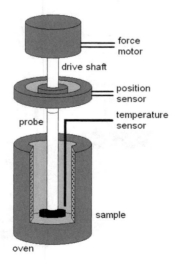

force motor

drive shaft

position sensor

temperature sensor

probe

sample

oven

Figure 5 *Schematic diagram of a thermomechanical analyser*

loudspeaker. Some form of proximity sensor measures the movement of the probe. This is usually achieved by using a linear variable differential transformer (LVDT) which consists of two coils of wire which form an electrical transformer when fed by an AC current. The core of the transformer is attached to the probe assembly and the coupling between the windings of the transformer is dependent upon the displacement of the probe. Other transducers such as capacitance sensors (which depend on the proximity of two plates – one fixed, the other moving) or optical encoders are used in certain instruments.

Most commercial instruments are supplied with a variety of probes for different applications (Figure 6). A probe with a flat contact area is commonly used for thermal expansion measurements where it is important to distribute the applied load over a wide area. Probes with sharp points or round-ended probes are employed for penetration measurements so as to determine the sample's softening temperature. Films and fibres, which are not self-supporting, can be measured in extension by clamping their free ends between two grips and applying sufficient tension to the specimen to prevent the sample buckling. Volumetric expansion can be determined using a piston and cylinder arrangement with the sample surrounded by an inert packing material such as alumina powder or silicone oil.

The equipment must be calibrated before use. The manufacturers, as well as various standardisation agencies, usually provide recommended

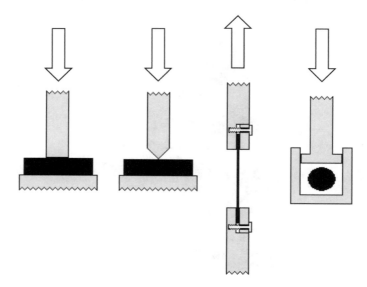

Figure 6   *TMA probe types (left–right): compression, penetration, tension, volumetric*

procedures. A full list of standard methods and calibration protocols for all thermal methods is given in the Appendix 3. Temperature calibration is usually carried out by preparing a sample that consists of a number of metal melting point standards, such as those used for differential scanning calorimetry, sandwiched between steel or ceramic discs. The melting of each standard causes a change in height of the stack as each metal melts and flows (Figure 7). Force calibration is often performed by balancing the force generated by the electromagnetic motor against a certified weight added to the drive train. Length calibration can be more difficult to carry out. A common check on the performance of the instrument is to measure the thermal expansion of a material whose values are accurately known (such as aluminium or copper).

Dynamic Mechanical Analysis

The distinction between a thermomechanical analyser and a dynamic mechanical analyser is blurred nowadays since many instruments can perform TMA-type experiments. The configuration of a DMA is essentially the same as the TMA shown in Figure 5 with the addition of extra electronics to apply an oscillating load and the ability to resolve the resulting specimen deformation into in-phase and out-of-phase components so as to determine E', E'' and tan δ. The facility for sub-ambient

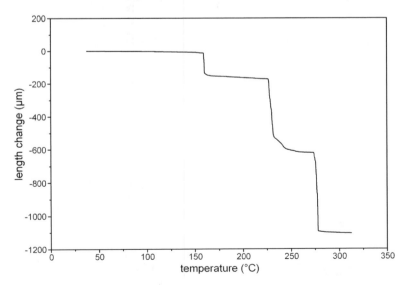

Figure 7 *TMA temperature calibration using indium (156.6 °C), tin (232.0 °C) and bismuth (271.4 °C). Heating rate: 5 °C min^{-1}, force 1 N, static air atmosphere*

operation is more common on a DMA than a TMA. The same recommendations about modest rates of temperature change are even more important for the larger samples used in DMA. Stepwise-isothermal measurements are often carried out for multiple frequency operation. In this experiment the oven temperature is changed in small increments and the sample allowed to come to thermal equilibrium before the measurements are made. The frequency range over which the mechanical stress can be applied commonly covers 0.01 to 100 Hz. The lower limit is determined by the amount of time that it takes to cover enough cycles to attain reasonable resolution of tan δ (10 s for one measurement at 0.01 Hz – though normally some form of data averaging is applied meaning that a measurement at this frequency can take a minute or more). The upper limit is usually determined by the mechanical properties of the drive system and clamps.

Different clamping geometries are used to accommodate particular specimens (Figure 8). Single or dual cantilever bending modes are the most common for materials which can be formed into bars. Shear measurements are used for soft, thick samples. Films and fibres are usually mounted in tension with loading arranged so that the sample is

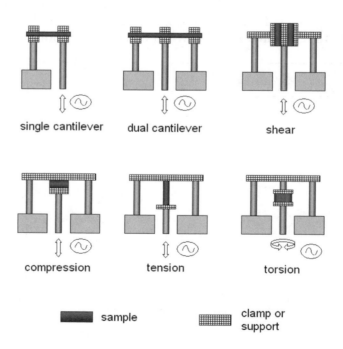

single cantilever   dual cantilever   shear

compression   tension   torsion

sample   clamp or support

Figure 8 *Common clamping geometries for dynamic mechanical analysis* (cf. *Figure* 1)

always in tension. Torsion measurements are normally done with a special design of instrument since most DMA's can only exert a linear rather than a rotational force.

The effect of temperature on the mechanical properties of a liquid can be investigated using a special type of dynamic mechanical analyser called an oscillatory rheometer. In this instrument the sample is contained as a thin film between two parallel plates. One of the plates is fixed while the other rotates back and forth so as to subject the liquid to a shearing motion. It is possible to calculate the shear modulus from the amplitude of the rotation and the resistance of the sample to deformation. Because the test is performed in oscillation, it is possible to separate the shear modulus (G) into storage (G') and loss modulus (G'') by measuring the phase lag between the applied strain and measured stress. Other geometries such as concentric cylinders or cone and plate are often used depending on the viscosity of the sample.

An alternative method for examining the dynamic mechanical properties of liquids is to coat them onto an inert support (typically a glass fibre braid). This measurement is termed Torsional Braid Analysis and does not provide quantitative modulus measurements since it is difficult to decouple the response of the substrate from that of the sample.

The method of calibration of DMA's varies from instrument to instrument and it is essential to follow manufacturer's recommendations. Temperature calibration can sometimes be done as for TMA's since many instruments can operate in this mode. Load or force calibration is often carried out using weights. It is difficult to achieve the same degree of accuracy and precision in modulus measurements from a DMA as might be obtained by using an extensometer without taking great care to eliminate clamping effects and the influence of instrument compliance (which can be estimated by measuring the stiffness of a steel beam). Extensometers are much bigger instruments and the size of test specimens is correspondingly larger. Additionally, they often only operate at room temperature. For many applications the user is, however, mainly interested in the temperatures at which changes in mechanical properties occur and the relative value of a material's properties over a broad range of temperatures.

Dielectric Techniques

A schematic diagram of a typical instrument is shown in Figure 9. The sample is presented as a thin film, typically no more than 1 or 2 mm thick, between two parallel plates so as to form a simple electrical capacitor. A grounded electrode surrounding one plate, known as a guard ring, is

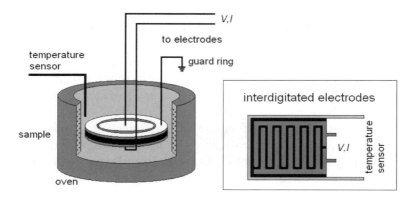

Figure 9   *Schematic diagram of a dielectric thermal analysis instrument. Inset shows a single-surface interdigitated electrode*

sometimes incorporated so as to improve performance by minimising stray electric fields. A thermocouple or platinum resistance thermometer is placed in contact with one of the plates (sometimes one on each plate) so as to measure the specimen's temperature. For specialised applications, such as remote sensing of large components, an interdigitated electrode is used (shown in the inset in Figure 9). These employ a pair of interlocking comb-like electrodes and often incorporate a temperature sensor (resistance thermometer). They can be embedded in structures such as a thermosetting polymer composite and the dielectric properties of the material monitored while it is cured in an autoclave.

A usual part of the calibration protocol for DETA is to measure the dielectric properties of the empty dielectric cell so as to take into account stray capacitances arising from the leads which must be of coaxial construction. Temperature calibration can be done by measuring the melting transition of a crystalline low molecular weight organic crystal such as benzoic acid placed between the electrodes.

TYPICAL EXPERIMENTS

This section discusses the most common types of experiments performed using TMA, DMA, TSCA and DETA by way of introduction to some of the more advanced applications described later.

Thermomechanical Analysis

Thermomechanical measurements can be carried out on a wide range of solid samples. The most usual mode of measurement is either in compression (for self-supporting samples) or tension (for thin films and fibres). Some materials exhibit anisotropic behaviour (particularly films or crystals) in that changes in dimensions will differ depending upon which axis the measurements are performed.

Thermal Expansion Measurements and Softening Temperatures

Plots of the change in length of a sample of a silicone gum rubber are shown in Figure 10. Three experiments were carried out on the material with different applied forces.[1]

At zero force a change in slope of the curve can be seen around $-60\,^{\circ}C$ due to the sample undergoing a change from glassy to rubbery behaviour. At this temperature the polymer chains acquire additional degrees of mobility which is seen as an increase in thermal expansion coefficient. The glass transition temperature (T_g) can be defined by finding the intercept of tangents to the linear portions of the length *versus* temperature plot above and below this region. When a force is applied to the

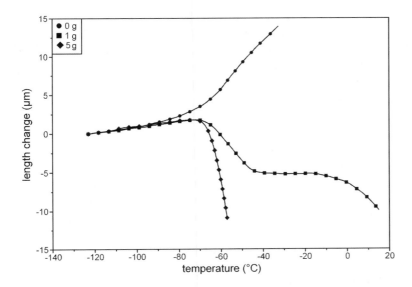

Figure 10 *Plots of change in length for a sample of silicone gum rubber under different applied loads on a flat-ended probe of 0.92 mm diameter. Heating rate 10°C min^{-1} under nitrogen, initial sample thickness 2.5 mm in all experiments*

specimen the probe deforms the material in inverse proportion to its stiffness. Below T_g the polymer is rigid and is able to resist the applied force, therefore its deformation is negligible. Above T_g the polymer becomes soft and the probe penetrates into the specimen. The temperature at which this occurs is called the materials' softening temperature and is highly dependent on the force applied to the sample.

Measurements of thermal expansion coefficients are useful in assessing the compatibility of different materials for fabrication into components. Mismatches in behaviour can cause stresses to build up when temperature changes occur resulting in eventual weakening and failure of the structure. Many crystalline materials can exist in a number of polymorphic forms which are stable at different temperatures. The transition between crystal structures is usually accompanied by a change in density and thermal expansion coefficient which can be detected by TMA.

Supporting information from differential scanning calorimetry is often useful in interpreting information from TMA – particularly when softening point determinations are made – since loss of mechanical integrity can occur due to melting, which gives an endothermic peak in DSC, or a glass–rubber transition, which causes a step change in heat capacity.

Dynamic Force Thermomechanical Analysis

Half way between conventional thermomechanical analysis and dynamic mechanical analysis is the technique of dynamic force (or load) TMA. This method uses a standard TMA instrument but the force is changed between two values in a stepwise (or sometimes sinusoidal) fashion. The dimensional changes of the specimen are monitored as a function of time (and temperature) but no attempt is made to determine the modulus and damping properties of the material.

An example of this is shown in Figure 11 for a carbon fibre reinforced epoxy resin composite beam measured in three point bending mode. The force on the probe was changed between 0.5 and 1.5 N and back every 12 s during the measurements. The average position of the sample deflection corresponds to a conventional TMA experiment under a force of 1 N. Above 90 °C the epoxy resin undergoes a glass–rubber transition and the specimen begins to deform under the applied load. The peak to peak amplitude of the probe movement is proportional to the compliance (= 1/stiffness) of the sample. This confirms that the sample is softer (more compliant) above T_g. It is possible to calculate the complex modulus of the material from the geometry of the sample, configuration of the test, applied forces and change in dimensions,[2] although such measurements are best carried out using a dynamic mechanical analyser (where this is

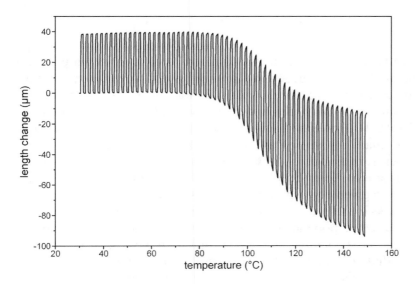

Figure 11 *Dynamic load thermomechanical analysis of a carbon fibre/epoxy resin beam measured in three point bending mode. The force steps between 0.5 and 1.5 N and back every 12 s. Heating rate 5 °C min^{-1} in air*

done automatically).

Some instruments are able to change the force on the sample during measurement so as to generate force–displacement curves in a manner similar to a conventional extensometer with the additional advantage of good control of specimen temperature. As the stress on the specimen is increased the material may creep under the applied load. When the force is removed the sample may attempt to recover its original dimensions (stress relaxation). Such tests are useful is assessing the resilience of materials such as rubber gaskets, O-rings and the like. This behaviour is related to the time and temperature dependent viscoelastic properties of the material discussed in the next section.

Dynamic Mechanical Analysis

Dynamic mechanical analysis is routinely used to investigate the morphology of polymers, composites and other materials. The technique can be particularly sensitive to low energy transitions which are not readily observed by differential scanning calorimetry. Many of these processes are time-dependent, and by using a range of mechanical deformation frequencies the kinetic nature of these processes can be investigated.

Single Frequency Temperature Scans

The most common DMA experiment is simply to measure the storage modulus (E') and mechanical damping factor (tan δ) against temperature at a single oscillation frequency. An example of this type of measurement is shown in Figure 12 for a specimen of the aliphatic polyester, poly(caprolactone). This polymer is typically highly crystalline and melts around 50–60 °C. However, the sample is not completely crystalline and contains a small amount of amorphous material which undergoes a glass–rubber transition at − 40 °C. For DMA and DETA work the glass transition is often called the alpha (α) transition and all lower temperature transitions are given corresponding Greek symbols (beta (β), gamma (γ) *etc.*). The peaks in tan δ at − 85 °C and − 130 °C correspond to the β and γ transitions in the polymer respectively and are due to the motion of short lengths of the polymer backbone rather than the large scale increase in mobility that accompanies the glass–rubber transition. It is very difficult to measure this type of behaviour by DSC, but the size and position of these transitions are often very important for a polymer's impact properties since they provide a means of dissipating mechanical energy as heat.[3]

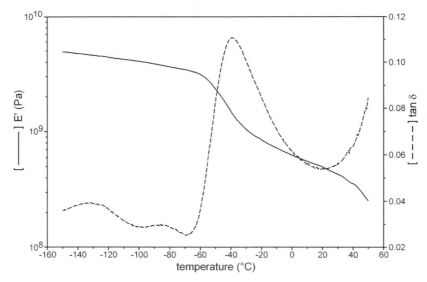

Figure 12   *DMA curve of poly(caprolactone) obtained in single cantilever bending at 1 Hz. Heating rate 2 °C min⁻¹ in air*

Step-wise Isothermal Frequency Scans

Figure 13 shows plots of storage modulus and damping factor for poly(ethylene terephthalate) (PET) film against temperature. The measurements were performed at a number of mechanical oscillation frequencies (0.3–30 Hz) and were carried out in a step-wise isothermal fashion with the temperature of the oven being raised by 5 °C and allowed to come to equilibrium before each frequency was successively applied to the specimen. It can be seen that the peak in tan δ moves to a progressively higher temperature as the measurement frequency is increased as a consequence of the time dependence of the glass–rubber transition. This maximum reflects the temperature at which the material can deform within the same time frame as the mechanical oscillation. Below this temperature the material is too sluggish to react to the deformation and behaves as a solid, above this temperature the material relaxes faster than the deformation and behaves as a viscous liquid.

Time–Temperature Superposition

It is well established that the observed temperature at which the glass–rubber transition occurs depends upon the time scale over which

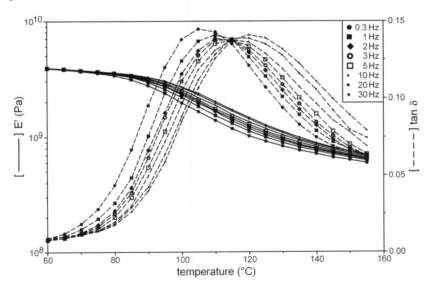

Figure 13 *DMA results for poly(ethylene terephthalate) film measured in tension at the different frequencies shown. The measurements were performed isothermally in 5 °C increments and the apparatus allowed to come to thermal equilibrium for 5 min before the sequence of measurements was performed*

one investigates molecular mobility (this applies to all methods of determining this parameter). To a first approximation the process can be treated as a simple thermally activated effect and the relationship between the temperature of maximum mechanical damping (T_{peak}) and the time scale (or frequency (f)) of the applied forcing variable (in this case mechanical deformation) can be analysed using a simple Arrhenius expression:

$$\ln(f) = \ln(A) - E_a/(RT_{peak}) \tag{16}$$

where E_a is the apparent activation energy for the process and R is the gas constant.

A more rigorous approach recognises that the glass–rubber transition is a co-operative effect and does not conform to the simple model described above. A common method for treating such a response is to superimpose data collected at different temperatures and frequencies onto one smooth curve. In Figure 14 the data from Figure 13 are presented on a frequency axis. It can be seen that if a curve at one temperature is chosen as a reference point then data from other temperatures can shifted in frequency to produce a smooth continuous change in properties spanning a wide frequency range. Ideally both the moduli and

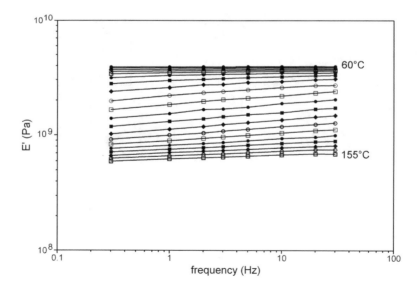

Figure 14 *Storage modulus data from Figure 13 shown as a function of frequency at different temperatures as indicated*

damping factor data should produce good overlays (Figure 15). The relationship between the frequency shift (a) at a specific temperature (T) and the reference temperature (T_{ref}) is usually expressed in terms of the Williams–Landel–Ferry (WLF) equation:[4]

$$\log[a(T)] = C_1(T - T_{ref})/(C_2 + T - T_{ref}) \tag{17}$$

where C_1 and C_2 are constants.

Time–temperature superposition is a means of extending the frequency range of dynamic mechanical data and has applications for the evaluation of materials for acoustic damping properties.

Dielectric Techniques

Measurements of electrical properties are particularly sensitive probes for the mobility of ions and dipoles within a specimen. Even non-polar materials like polyethylene often contain polar impurities which give sufficient response for the behaviour of the specimen to be analysed by these methods. The typical experiments described below illustrate the specialised niche occupied by TSCA and DETA in that the measurements are placed in the context of addressing a particular problem. Applications

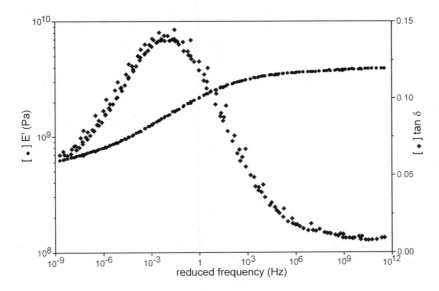

Figure 15 *"Master curve" of data from Figure 13 with a reference temperature of 100°C*

of DETA to remote sensing of a sample's properties are described later.

Thermally Stimulated Current Analysis

The stability of lyophilised products is crucial in pharmaceutical industry. Freeze-drying a drug formulation is a common way to preserve the active component during the storage until its final use. The active ingredient is diluted and embedded in an excipient formulation during the freeze-drying process. The physical properties of the excipient in the solid state dictate the stability behaviour of the formulation. Knowledge of the glass transition (T_g) and melting (T_m) temperatures are essential to predict shelf life. If the lyophilised product is stored above one or the other transition, it is likely that the material will change over time and the stability of the active ingredient cannot be guaranteed. A knowledge of T_g is also very important for the lyophilisation itself. In particular, the interaction of the system with water is essential to model the freeze-drying process.

DSC is the most appropriate technique to give the best information concerning melting. The data collected are thermodynamic variables such as heat capacity or enthalpy of fusion, *etc.* However, the change in heat capacity that occurs at the glass transition can often be rather small (particularly for materials with low amorphous content) and there may not be enough sensitivity to detect the glass transition without ambiguity. A feature of the thermally stimulated current technique is that the current flow is directly proportional to the strength of the electric field. Thus it is possible to "magnify" weak transitions by increasing the polarisation voltage.

Figure 16 shows the results from a TSCA of a freeze-dried drug. In essence, two experiments are shown. The lower curve represents the current flow during polarisation under an applied field of 100 V as the sample was heated from 20 °C at 7 °C min^{-1} to 70 °C. The sample was then cooled to -10 °C and the electric field switched off. The upper curve shows the discharge of the sample as it was heated to 90 °C at the same rate in the absence of any external field. The peaks in both curves represent the glass transition of the drug. Note that the peak in the polarisation curve occurs at a higher temperature than the discharge curve which is typical of this technique. The discharge curve is known as the "global" discharge curve – a more advanced technique can be used to carry out the polarisation experiment over a narrow temperature window in order to examine the time and temperature dependence of the discharge process.

TSCA is more often applied to polymeric materials,[5] but this example

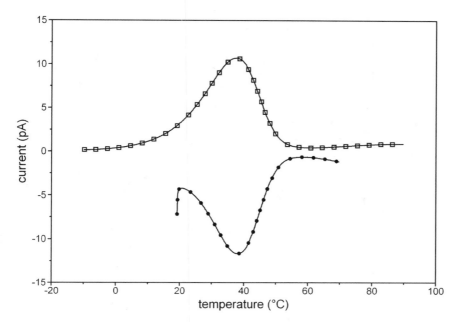

Figure 16 *Thermally stimulated current analysis of a freeze-dried drug. Lower curve: polarisation current. Upper curve: depolarisation current. Heating rate: 7°C min⁻¹ for both experiments*

illustrates the use of the technique on a system where the detection of small amounts of amorphous material is of critical importance to its stability.

Dielectric Thermal Analysis

Measurement of thermal transitions of thin films can be problematic due to the small amount of material that is present. The glass transition temperature of adhesives affects the tack and bonding behaviour, but analysis by DSC can be difficult unless the material is physically removed from any backing. This can be aided by swelling the adhesive in solvent and scraping it off the substrate with a blade. There is a danger that the properties of the sample could be changed by this process even if the adhesive is subsequently dried so as to remove any residual solvent. Dielectric measurements, however, are ideally suited to testing thin films and, so long as the backing undergoes no thermal transitions in the region of interest, the adhesive can be measured *in situ* without any preparation.

Figure 17 shows plots of ε' and ε'' obtained at various frequencies for a

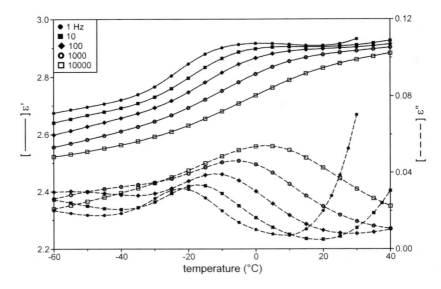

Figure 17   *Dielectric analysis of a thin layer of pressure sensitive adhesive on poly(ethylene terephthalate) film at the different electrical field frequencies shown. Heating rate: 2°C min⁻¹*

sample of poly(ethylene terephthalate) film coated with a 50 μm thick layer of pressure sensitive adhesive. The glass–rubber transition of the adhesive can be observed as a step increase in ε' and a peak in ε'' which shifts to higher temperatures as the measurement frequency increases. Figure 18 shows the response of the base film itself. The PET has a weak transition due to local movement of dipoles associated with its β transition which has a much stronger frequency dependence than that of the adhesive. This causes some uncertainty with the assignment of T_g of the adhesive at 10 kHz where there is expected to be some overlap. At lower frequencies the peak in ε'' due to the substrate is well separated from the T_g of the adhesive and the latter can be measured more accurately. Measurement of the properties of paints and other coatings can be carried out in the same way. In some cases the film may be put down directly onto a metal substrate which can then be used as one of the electrodes.

   The principles of time–temperature superposition can be used with equal success for dielectric measurements as well as dynamic mechanical tests. Analysis of the frequency dependence of the glass transition of the adhesive in the system described above shows that it follows a WLF type dependence whereas the β transition of PET obeys Arrhenius behaviour. This type of study can be used to distinguish between different types of relaxation phenomena in materials.

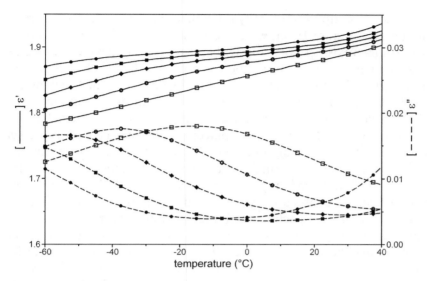

Figure 18 *Dielectric analysis of uncoated poly(ethylene terephthalate) film under the same conditions as Figure 17*

APPLICATIONS

The following examples are used to demonstrate the broad range of applications of thermomechanical and thermoelectrical measurements. Many of the applications are not typical of the routine types of tests described earlier, but are placed here to show the diversity of these techniques in characterising materials and provide short "case studies" which present the methods in the context of addressing particular problems.

Thermomechanical Analysis

Whilst thermomechanical measurements are routinely used to investigate mechanical stability and measure thermal expansion coefficients, two examples are given which illustrate the use of the technique on the small scale as a micro-analytical tool to identify the distribution of two materials within a matrix and also on a larger scale to investigate the firing of a ceramic material.

Localised Thermomechanical Analysis of Pharmaceuticals

Micro-thermal analysis is discussed further in Chapter 6. Here the technique is employed to illustrate the use of thermomechanical measurements on small samples to identify different components in a mixture of two materials.[6] Figure 19 shows a micrograph of the surface of a tablet pressed from a mixture of benzoic acid and salicylic acid which is used as a test specimen for dissolution studies to mimic the effect of digestion within the gut. The image was obtained by using an atomic force microscope which uses a fine stylus to measure the height of the sample as the tip of the stylus is scanned over the surface of the specimen. Having obtained the image, the tip may be placed on the sample with a pre-set force and heated by passing a current through it. At the same time its vertical displacement is measured. In this way, thermomechanical analysis of areas less than 5 μm^2 may be carried out. Several different locations may be measured consecutively and used to "map" the softening or melting behaviour of a specimen with high spatial resolution.

Figure 20 shows the results of carrying out such measurements at different points marked in Figure 19. It is evident that the material examined at points 1 and 3 has a lower melting temperature than the other locations. Since benzoic acid melts at 122 °C and salicylic acid melts at 159 °C, the results from such measurements are sufficient to identify the distribution of these materials within the image.

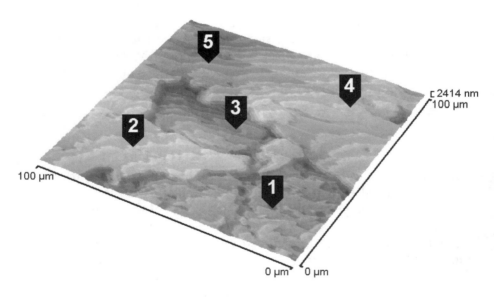

Figure 19   *Atomic force micrograph of the surface of a pressed compact tablet containing a mixture of benzoic acid and salicylic acid. Labels indicate positions of localised thermomechanical analysis in Figure 20*

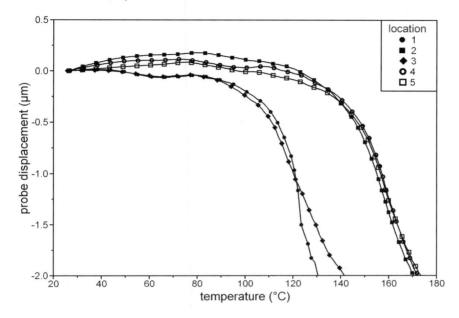

Figure 20 *Localised thermomechanical analysis of different areas shown in Figure 19. Heating rate $10\,^{\circ}C\ s^{-1}$*

Rate Controlled Sintering of Ceramics

The compaction and sintering of high temperature refractory materials may be studied by thermomechanical analysis. In many cases it is desirable that the ceramic changes in dimensions in a uniform manner. To achieve this the rate of heating can be controlled by the rate of change of dimensions of the specimen. This can be done by heating the sample at a fixed heating rate and then stopping heating when the rate of change of length exceeds a certain limit. The process is allowed to continue isothermally until the rate falls below the limit and then heating is recommenced. An alternative approach is illustrated in Figure 21. Here, heating takes place at a maximum fixed rate of temperature rise ($2\,^{\circ}C$ min^{-1}) until the rate of shrinkage reaches a pre-set value. Thereupon the rate of temperature change is controlled so as to maintain a constant rate of shrinkage. The desired rate of shrinkage may be changed during the experiment between different values (in this case stepped alternately between 1 and $2\ \mu m$ min^{-1} during the temperature rise) so as to explore the relationship between the sintering kinetics and temperature.[7]

A similar technique has been described for DMA whereby the temperature program was controlled by constraining the rate of change of mech-

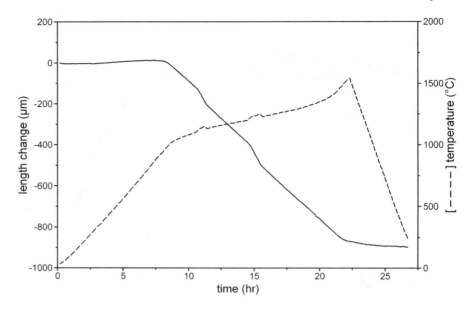

Figure 21   *Rate controlled sintering of a ceramic under air. The rate of temperature rise is controlled by the rate of change in length over pre-set temperature ranges with a limiting maximum heating rate $2\,^{\circ}C\ min^{-1}$*

anical properties (*e.g.* storage modulus) to within certain limits. This approach was shown to be effective in resolving the multiple glass transitions of a polymer blend.[8]

Dynamic Mechanical Analysis

Dynamic mechanical measurements are not limited to running experiments on samples in air or inert gases. With care, measurements can be carried out with the test specimen immersed in a liquid or on liquid samples themselves as the following examples demonstrate.

Dynamic Mechanical Analysis of Fibres under Dye Bath Conditions

The dyeing and washing behaviour of regenerated and synthetic fibres are markedly dependent upon temperature. For example, acrylic fibres must be dyed above T_g in order to facilitate dye diffusion; in contrast, the characteristic high wet fastness properties of the resultant dyeings can be attributed primarily to the relative absence of dye diffusion that results from such aqueous treatments (*e.g.* laundering) being carried out at temperatures below the T_g of the fibre. Although the effect of temperature on the physical properties of dry fibres is readily performed using conventional thermoanalytical techniques (*e.g.* DSC), it is less easy to examine the behaviour of fibres under dye bath conditions owing to the presence of water. Of particular interest is the effect of "carriers" – additives to the dye bath which are used to accelerate the rate of dye diffusion within the hydrophobic fibres. The mechanism by which these materials enhance dye uptake is thought to be by plasticising the fibre by reducing its glass–rubber transition temperature and thus increasing the segmental mobility of the polymer chains.

Using a dynamic mechanical analyser, it is possible to carry out experiments with the sample immersed in a liquid. For this type of measurement, the instrument is fitted with a metal liner which is inserted into the oven so that the sample and clamps can be immersed and the temperature of the bath programmed in the usual way. Results from testing acrylic fibres under different conditions are shown in Figure 22.

It can be seen that the peak damping factor is reduced from around 90 to 72 °C by the presence of water. Addition of benzyl alcohol to the water bath (as a model "carrier") further depresses and broadens the peak in tan

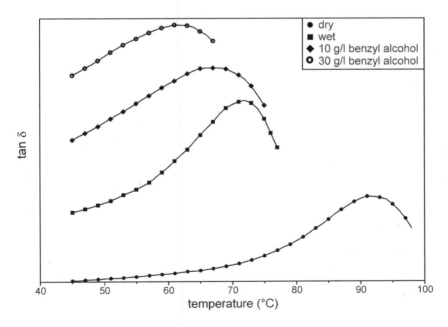

Figure 22 *Dynamic mechanical analysis of acrylic fibres in air, water and solutions of benzyl alcohol. Measurements performed in tension at a frequency of 1 Hz, heating rate 1 °C min⁻¹. Curves shown offset for clarity*

δ in a fashion typical of the action of a plasticiser thus confirming the nature of carrier activity.[9]

Variable Temperature Oscillatory Rheometry of Food Additives

Cellulose ethers, such as methyl cellulose, are widely used as thickening agents in a variety of foodstuffs such as pie fillings and potato croquettes. These substances possess the unusual feature of forming a reversible gel structure on heating which serves to maintain the mechanical integrity of the product during baking. On cooling, the gel structure breaks down and the original texture of the mixture is regained.

Measurements on a 10% solution of methyl cellulose in water were carried out using an oscillatory rheometer. Plots of storage and loss moduli (G' and G'' respectively) against temperature are shown in Figure 23. Below 55 °C the loss modulus is higher than the storage modulus, indicating that the specimen is responding more like a liquid than a solid. Above this temperature, the situation is reversed and the specimen has predominantly solid-like characteristics. The cross-over between G' and G'' corresponds to the formation of a cross-linked gel network and the

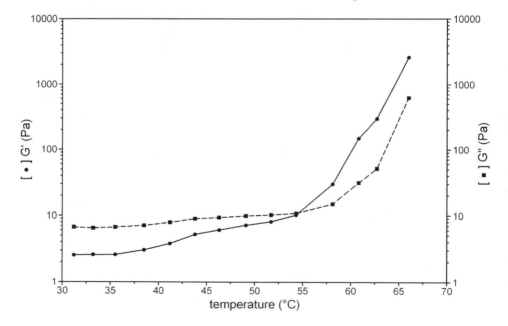

Figure 23   *Storage (G') and loss shear (G'') modulus for a 10% solution of methyl cellulose in water as a function of temperature. Measurement frequency 1 Hz, heating in steps with 5 min dwell time between steps*

transition between the two types of behaviour.[10] The effect of different additives on gel formation can be a useful indicator of synergistic (lowering of gel temperature) or antagonistic (raising of gel temperature) interactions between materials.

DMA is regularly used to study the chemical reactions which lead to cross-linking of thermosetting resins, such as those used in the manufacture of composites. At high degrees of network formation the rubbery cross-linked gel will vitrify into hard glassy material and the storage modulus will increase by several orders of magnitude. Plotting gel point and vitrification point against temperature and time leads to a time–temperature–transformation (or Gillham–Enns[11]) diagram which can be used to map out the curing of thermosetting polymers. The use of dielectric analysis in the following section will show how this technique can also be used to follow chemical cross-linking and the reverse effect of breakdown of molecular structure brought about by exposure to UV radiation.

Dielectric Thermal Analysis

Dielectric thermal analysis involves monitoring the viscosity of a system *via* its ability to store or transport electrical charge. Changes in the degree of alignment of dipoles and the ion mobility provide information pertaining to physical transitions in the material and to material properties such as viscosity, rigidity, reaction rate and cure state. By use of remote dielectric sensors, the measurements can be made in actual processing environments such as presses, autoclaves, and ovens.

In Situ Cure Monitoring of Composites

In dielectric cure monitoring, the ion mobility (electrical conductivity) of the material is of greatest interest. Almost all materials contain current carriers, which are charged atoms or charged molecular complexes. The application of a voltage between a set of electrodes will create an electric field that forces those ions to move from one electrode towards the other. Ions encounter something analogous to viscous drag as they flow through a medium filled with molecules, and their mobility through this medium determines the conductivity. Conductivity is inversely proportional to viscosity. Ions moving through very fluid, watery materials have high mobility and conductivity – resulting in low resistivity that correlates with low viscosity. Conversely ions moving through very rubbery materials have low mobility and conductivity corresponding to the high viscosity. It is important to note that beyond some point in the cure the

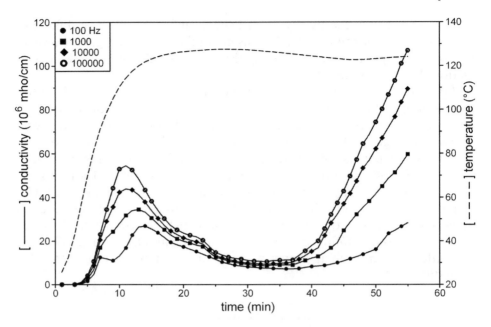

Figure 24   *Plots of ionic conductivity against time during the cure cycle of a glass fibre reinforced epoxy resin panel using an interdigitated sensor. Temperature profile measured by a resistance thermometer in the sensor*

physical viscosity will climb so high that it is no longer measurable, even though the cross-linking reaction has not reached completion. Because increasing polymerisation continues to affect ionic motion, dielectric measurements retain sensitivity past the time when ionic and physical viscosity deviate. Consequently, with proper interpretation, dielectric measurements are useful throughout the entire curing process for determining changes in viscosity and rigidity, and are extremely sensitive in determining the end of cure.

Figure 24 shows the conductivity of a glass fibre reinforced epoxy resin composite during curing in a heated press. An interdigitated electrode was embedded in the sample during lay-up of the specimen, and a resistance thermometer in the electrode monitors the temperature of the sample directly during processing. Initially the conductivity of the specimen increases as it is heated to the cure temperature and the resin becomes less viscous. Cross-linking causes the viscosity to increase and, as a consequence, the conductivity decreases. Towards the end of the reaction the material forms a highly cross-linked network and shows a strong frequency dependence in conductivity. Dielectric measurements readily lend themselves to being carried out simultaneously with dynamic mechanical analysis when experiments are performed in compression or torsion. The sample is usually mounted between parallel plates which are used to apply the mechanical stress – electrical connections can be established to these and used to make a dielectric measuring cell.[12]

In Situ Monitoring of the UV Degradation of Adhesives

Tinted self-adhesive plastic films are a popular means of limiting the effects of sunlight on the interiors of buildings and vehicles. Stuck to the inside surfaces of windows, these are used to filter infra-red and ultraviolet (UV) radiation, thereby avoiding heat build-up in offices and vehicles, and fading of interior upholstery. After the glass itself, the adhesive is first to receive solar radiation and must be stabilised against photodegradation by an appropriate choice of polymer and stabiliser package. This not only protects the adhesive but, by blocking short wavelength radiation, suppresses fading of the dyes used to colour the film.

The performance of candidate systems is assessed by exposure to intense radiation equivalent to the solar spectrum in a hot, humid environment during accelerated ageing. Measurements of the UV transmission of the film are made at regular intervals in addition to critical evaluation of the optical appearance of the plate by eye. Since these products are to be used in vehicles, any blemishes in the film brought about by degradation of the adhesive are extremely undesirable.

In such a study,[13] test panels of window film were mounted on glass plates for accelerated ageing. A small hole was cut in the plastic film and an interdigitated single surface dielectric sensor applied to the exposed surface of the adhesive after chilling the plate in a freezer to aid removal of the backing. Measurements of dielectric loss and permittivity were carried out from 0.1 to 100 kHz in decade steps at ambient temperature before and after 600 and 1200 h exposure.

Values of dielectric loss (ε'') at 0.1 Hz for the same adhesive containing three candidate stabiliser packages are given in Table 1 at different stages

Table 1 *Dielectric loss factor, ε'' (0.1 Hz) for adhesive with different stabiliser packages during accelerated weathering*

| | Stabiliser package | | |
Sample	"Standard"	"Poor"	"Good"
Initial	3.00	2.86	3.14
After 600 h	8.59	12.6	3.18
After 1200 h	10.4	71.0	6.45

of weathering. The data indicates that, of the three stabiliser formulations, the "poor" package showed the largest increase in ε''. UV transmission measurements showed that, although starting with the same value (*ca.* 0.4%), the transmission of the "standard" package rises to 5.7% compared to around 2.5% for both adhesives containing the "good" and "poor" formulations after 1200 h weathering. The "poor" sample exhibited severe optical distortion of the film, but remained light fast, the "good" sample (the same formulation as the "poor" sample but with an additional stabiliser) maintained good UV absorbtion and showed no blemishes. Even after 600 h exposure it was apparent that changes in adhesive behaviour could be detected by this technique before deterioration in appearance could be seen by eye.[13]

MODULATED TEMPERATURE THERMOMECHANICAL AND DIELECTRIC TECHNIQUES

In the same way that applying a temperature modulation to DSC can be used to separate thermally reversing thermal events (such as the glass transition) from thermally non-reversing ones (*e.g.* crystallisation and curing), the same principles can be applied to TMA. This approach enables one to separate reversible dimensional changes due to thermal expansion from irreversible effects such as creep or stress relaxation.[14] Modulated temperature DMA has been developed as a means of investigating the reversible melting of polymers.[15] A sinusoidal heating program has also been employed in TSCA to separate reversible pyroelectric currents from non-reversible thermally stimulated discharge of heated dielectric materials.[16]

CONCLUDING REMARKS

The description of thermomechanical and thermoelectrical measurements in a modest chapter such as this is an ambitious exercise. The author has attempted to cover a wide range of methods and applications with the intention of illustrating the diversity of this field whilst emphasising the relationships between the static techniques (such as TMA and TSCA) and the dynamic techniques (dynamic force TMA, DMA and DETA). With the exception of TMA, these methods are often promoted as some of the more "advanced" thermal analysis techniques. It is hoped that the preceding pages help to dispel this myth without belittling their ability to measure useful properties.

ACKNOWLEDGEMENTS

It is a pleasure to acknowledge the assistance of George Collins (Ther-Mold Partners), Huan Lee (Holometrix-Micromet), Kevin Menard (Perkin-Elmer Corporation) and O. Toft Sørensen (Risø National Laboratory) for their advice and provision of original data used within this chapter.

FURTHER READING

D. Q. M. Craig, *Dielectric Analysis of Pharmaceutical Systems*, Taylor and Francis, London, 1995.
J. Goodwin and R. W. Hughes; *Rheology for Chemists*: *An Introduction*, Royal Society of Chemistry, Cambridge, 2000.
J. D. Ferry, *Viscoelastic Properties of Polymers*, Wiley, New York, 3rd. edn., 1980.
N. G. McCrum, B. E. Read and G. Williams, *Anelastic and Dielectric Effects in Polymeric Solids*, Dover, New York, 1991.
K. P. Menard, *Dynamic Mechanical Analysis*: *A Practical Introduction to Techniques and Applications*, CRC Press, Boca Raton, FL, 1999.
L. E. Nielsen and R. F. Landel, *Mechanical Properties of Polymers and Composites*, Dekker, New York, 2nd edn., 1993.
A. T. Riga and C. M. Neag, (ed.), *Materials Characterization by Thermomechanical Analysis*, ASTM STP 1136, American Society for Testing and Materials, Philadelphia, 1991.
J. P. Runt and J. J. Fitzgerald (ed.), *Dielectric Spectroscopy of Polymeric Materials*: *Fundamentals and Applications*, American Chemical Society, Washington DC, 1997.
M. Reading and P. J. Haines, "Thermomechanical, dynamic mechanical and associated methods" in P. J. Haines (ed.), *Thermal Methods of Analysis*: *Principles, Applications and Problems*, Blackie, Glasgow, 1995, pp. 123–160.

REFERENCES

1. C. M. Earnest, "Assignment of Glass Transition Temperatures Using Thermomechanical Analysis", in R. J. Seyler (ed.), *Assignment of the Glass Transition*, ASTM STP 1249, American Society for Testing and Materials, Philadelphia, 1994, pp. 75–87.
2. R. Reisen, G. Widmann and R. Truttmann, *Thermochim. Acta*, 1996, **272**, 27.

3. L. Woo, S. P. Westphal, S. Shang and M. Y. K. Ling, *Thermochim. Acta*, 1996, **284**, 57.

4. M. L. Williams, R. F. Landel and J. D. Ferry, *J. Am. Chem. Soc.*, 1955, **77**, 3701.

5. J. R. Saffell, A. Matthiesen, R. McIntyre and J. P. Ibar, *Thermochim. Acta*, 1991, **192**, 243.

6. P. G. Royall, D. Q. M. Craig, D. M. Price, M. Reading and T. J. Lever, *Int. J. Pharm.*, 1999, **191**, 97.

7. O. Toft Sørensen, *Thermochim. Acta*, 1981, **50**, 163.

8. M. Reading, "Controlled Rate Thermal Analysis and Beyond", in *Thermal Analysis – Techniques & Applications*, ed. E. L. Charsley and S. B. Warrington, The Royal Society of Chemistry, Cambridge, 1992, p. 127.

9. D. M. Price, *Thermochim. Acta*, 1997, **294**, 127.

10. J. Desbrieres, M. Hirrien and S. B. Ross-Murphy, *Polymer*, 2000, **41**, 2451.

11. J. K. Gillham and J. B. Enns, *Trends Polym. Sci.*, 1994, **2**, 406.

12. B. Twombly and D. D. Shepard, *Instrum. Sci. Technol.*, 1994, **22**, 259.

13. D. M. Price, *J. Therm. Anal. Cal.*, 1997, **49**, 953.

14. D. M. Price, *Thermochim. Acta*, 2000, **357/358**, 23.

15. A. Wurm, M. Merzlykov and C. Schick, *Thermochim. Acta*, 1999, **330**, 121.

16. E. J. Sharp and L. E. Garn, *J. Appl. Phys.*, 1982, **53**, 8980.

Chapter 5

Calorimetry

R. J. Willson

GlaxoSmithKline, New Frontiers Science Park (South), Harlow, Essex, UK

INTRODUCTION

Where change can occur, change is inevitable. The Second Law of Thermodynamics may be stated as: "Everything naturally tends to its most stable state". Some materials benefit from change, becoming more useful consequently, but, for most materials, change represents a reduction in useful properties and loss in quality. In some cases, change can even result in a danger to health. For the myriad materials around us, change can be obvious; discolouring of paint, rusting of metal, decay, or a raging bush fire. Other changes can be subtle, such as a polymorphic change of a crystal, fatigue in a metal, or the surface sorption of vapour. In the majority of cases, change can have a considerable influence on the useful properties of a material. What appears to be the same substance by casual observation may have very different properties if change has occurred.

Industries that have a reliance on materials have a burdensome responsibility to ensure constant quality. Reliability comes from knowledge of the process of change and the predictions about change resulting from environmental conditions. Like freshly baked bread, most materials are at the highest quality at the time of manufacture but depreciate with time. Since it is important to understand change, effort and money are therefore invested in an attempt to ensure quantification of change. This provides the basis for a shelf life specification and recommendation for the conditions under which a material can be used.

There are many recognised techniques for evaluating the structure and consistency of materials, for example spectroscopic analysis. These are often used in authentication, and to ensure batch-to-batch uniformity. Calorimetry has found value for recognising subtle differences in materials that are not apparent using any other technique. In addition, calorimetry can quantify change in terms of rate, how much and the probability for a change to occur.

Change can be promoted by various means. Chemical modifications have been observed from oxidation and reduction, hydrolysis and photolytic reactions, reactions with excipients and with solvents and autocatalytic reactions. Change can also come about from modifications to the solid state. Common to crystalline materials are polymorphic changes, changes in hydrate or solvate forms, phase changes and change in surface area. As indicated in the first paragraph, change is inevitable if change is possible. However, the rate of change may be dependent on environmental conditions such as temperature, pressure, humidity, and mechanical action.

Definition of Calorimetry

Calorimetry is defined as the measurement of heat. It has been used to study reactive systems since 1780 when Lavoisier and Laplace first studied the respiration of a guinea pig in an ice calorimeter.[1] The quantity of water collected and the rate of melting gives a thermodynamic and kinetic evaluation for respiration. Since then, considerable progress has been made in the technology of calorimeters. Lavoisier was restricted to measuring exothermic reactions at 273.15 K and the sensitivity of the instrument was dependent on the accuracy of weighing the melted ice. Modern calorimeters can record, directly, exo or endothermic reactions with signals as low as 5×10^{-8} W, as shown in Figure 1. Such sensitivity permits greater specificity in the interpretation of calorimetric data than the ice calorimeter could achieve.

Instruments have been designed to measure energy changes from temperatures a little above absolute zero up to temperatures in excess of two or three thousand kelvin. Homemade instruments were the rule until about 1960, but even so, high quality measurements were common. Calorimetry was not a tool for routine use in the laboratory. Advances in electronics and the development of new designs have changed this situation so dramatically that calorimetry is now widely used for the day-to-day physical and chemical characterisation of materials within a wide range of research applications. This chapter will discuss a variety of non-scanning calorimeters, but will emphasise the potential of current commercially available instruments. Physical and chemical processes for

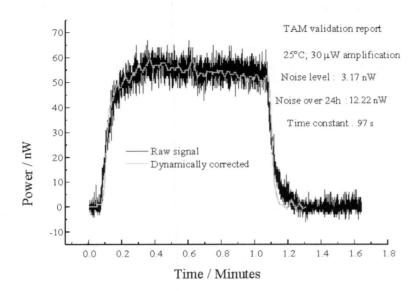

Figure 1 *A calorimetric signal showing the sensitivity of an isothermal microcalorimeter (Thermometric, TAM) fitted with nW amplifiers. The experiment was performed at 298.15 K using 3 cm^3 glass ampoules containing ~ 1 g of dry talcum powder. A 50 nW electrical input was applied for 1 min and then switched off*

solids, liquids and gases have been investigated and, since each application addresses specific problems, descriptions of many different types of calorimeter may be found in the literature.[2-6]

Modern calorimetry has found many useful applications in material development as described later. The high sensitivity and signal stability over long time periods make modern calorimeters very suitable for the study of change. Most instruments require small amounts of material, are very sensitive and, at least in principle, can record all changes that occur.

THERMOCHEMISTRY AND THERMODYNAMICS

To appreciate the usefulness of calorimetry it is important to recognise the principles of thermodynamics upon which it is based. Together with the laws of physics, these govern the chemical and physical changes that occur and through which the calorimetric instruments operate. A full account of the development and mathematics of thermodynamics is to be found in the many excellent textbooks.[7-9] Brief summaries of the principles are given here.

The First Law of Thermodynamics: The Energy of an Isolated System is Constant

Another expression of this law considers the internal energy, U, of a system and how it is effected by the work done on it (w) and heat energy added to it (q). The change in internal energy of a system, ΔU, may be written as

$$\Delta U = q + w \tag{1}$$

If heat is added to a system, the temperature rises. If the system is at constant volume, no work is done, so for an infinitesimal change:

$$dU = = dq_V = C_V dT \tag{2}$$

where, C_V is the heat capacity at constant volume. This applies when combustion takes place in a closed bomb calorimeter. At constant pressure, enthalpy (H) is defined to take into account the work done on the system by the pressure:

$$H = U + PV \tag{3}$$

Heat added to raise the temperature at constant pressure is given as

$$dH = dq_p = C_p dT \tag{4}$$

where C_p is the heat capacity at constant pressure. The heat capacities may not be constant with temperature. Applying an electrical power for a measured time allows the instrument to be calibrated. The temperature rise is a consequence of the applied power and can be determined from the current I passed through a resistance R for a time t as shown in Figure 1:

$$Pt = I^2 Rt \tag{5}$$

When a change occurs with the evolution of heat to the surroundings at constant pressure it is referred as exothermic and the enthalpy of the system decreases (ΔH is negative). If the change takes in heat from the surroundings it is endothermic (ΔH is positive). Many reactions have been studied and the value of the enthalpy of reaction determined accurately. Tables of standard molar enthalpies of formation $\Delta_f H$ at 25°C are to be found in several sources[10,11] and the standard enthalpy change for another reaction may be calculated from them using Hess's Law. This states that "the enthalpy change of a reaction may be expressed as the sum of the enthalpy changes into which the overall reaction may be divided".[8] Hess's law allows the calculation of enthalpy changes for reactions which cannot be measured directly, either because they occur too slowly for the instrument to follow, or because they do not occur naturally.

The Second Law of Thermodynamics: In an Isolated System, Spontaneous Processes Occur in the Direction of Increasing Entropy

The entropy (S) of a system is a thermodynamic function related to the statistical distribution of energy within that system. The Second Law means that the system and surroundings will change spontaneously from a state of low probability to a state of maximum probability, which will be the equilibrium state. One example of this is the mixing of two ideal gases at atmospheric pressure. This will occur spontaneously if a barrier between them is removed. No heat or pressure change is involved, but the change clearly involves an increase in the randomness of the system as the gases become mixed.

The Third Law of Thermodynamics: All Perfect Crystals Have the Same Entropy at Absolute Zero

To use entropy as a criterion of spontaneous change, it is necessary to investigate both system and surroundings. For that reason, a further thermodynamic function, the free energy, or Gibbs function G, is used. This combines the enthalpy H and the entropy S and allows the combination of the effects of both H and S on the system only, generally at constant pressure. The quantities U, H, G and S are referred to as state functions, since they depend only on the state of the system.

$$G = H - TS \qquad (6)$$

The free energy change (ΔG) serves as a good indicator for the potential for a reaction to take place and how far a reaction will go towards completion. Under standard conditions, $\Delta G^{\ominus}$ is related to the equilibrium constant (K_{eq}) of a reaction at a given temperature:

$$\Delta G^{\ominus} = - RT \ln(K_{eq}) \qquad (7)$$

All reactions may be written as equilibrium processes. Some equilibria lie far to the right, indicating that the majority of the reactants form products. Some lie far to the left indicating the majority of reactants remain as reactants. Similarly, some lay between these extremes. The International Conference on Harmonisation (ICH)[12] of technical requirements for registration of pharmaceuticals publish guidelines recommending the reporting threshold for degradation products in pharmaceutical materials. The amount of degradation product allowed before the requirement of degradant identification is currently 0.1% for a high dose drug product. A reactant product quantity of 0.1%, at equilibrium, relates to an equilibrium constant (K_{eq}) of 1×10^{-3}. Beezer *et al.*[13] have shown that the calculation of K_{eq} is possible from isothermal calorimetric data where enthalpy changes for a reaction are determined as a function of temperature. From equation (7), an equilibrium constant of 1×10^{-3} corresponds to a $\Delta G^{\ominus}$ of $+ 17.12$ kJ mol^{-1}, at 298.15 K. Most spontaneous reactions (meaning most natural reactions) will result in a decrease in Gibbs energy and an increase in the overall entropy.

For example, crystallisation from a saturated solution is a favourable process with a negative ΔG. This is because the process is exothermic (ΔH negative) and entropically favourable (ΔS positive). The removal of solute from the solvent has an associated entropy gain greater than the decrease in entropy from building of the crystal lattice. The calculation of an equilibrium constant can provide a powerful investigative tool for comparing materials. For example, the equilibrium solubility of polymorphs can be studied as a function of temperature. From which a phase diagram can be made, *e.g.*, a plot of ΔG against temperature. This can give considerable insight into how a material would behave under different environmental conditions and monotropic and enantiotropic behaviour identified, as shown in Figure 2.

By measuring the equilibrium constant at several temperatures, a further equation may be derived that allows the enthalpy change to be calculated for the reaction from the van't Hoff equation:

$$d(\ln K)/dT = \Delta H^{\ominus} / (RT^2) \tag{8}$$

One further relationship, which may be used for reactions that take place in an electrochemical cell, is that between the free energy and the electromotive force (E) of the cell:

$$\Delta G = - nFE \tag{9}$$

Measuring the emf of the cell at several different temperatures allows the calculation of the enthalpy change, giving another indirect measurement technique.

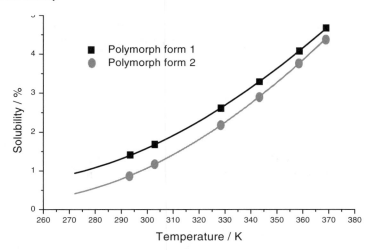

Figure 2 *Calorimetry can be useful for examining relative polymorphic stability by comparing the solubility of the forms. Solubility can be derived from the heats of
solution for a polymorphic form dispersed into a solvent divided by the enthalpy
change of fusion ($\Delta_{fus}H$). A correction can be made for the enthalpy of solvation
using the amorphous form of the material. The graph shows the relative solubility
of two polymorphs of a drug substance in development with GlaxoSmithKline*

In addition to the thermodynamics, calorimetric measurements rely
upon the laws of heat transfer. Transfer of heat by radiation is most
important at high temperatures, and is minimised by using bright, clean
metal surfaces. Conduction will occur through the material of the
calorimeter, but not through a vacuum. The conduction law for heat
transfer across a sample of area A, thickness dx is:

$$\frac{dq}{dt} = -kA\left(\frac{dT}{dx}\right) \tag{10}$$

Here, dq/dt is the heat transferred per unit time, dT/dx is the temperature
gradient in the $+x$ direction and k is the thermal conductivity of the
sample. On occasion, the geometric constants and thermal conductivity
may be combined into a heat transfer coefficient h, giving the heat
conduction from sample to environment:

$$\frac{dq}{dt} = h(T_S - T_E) \tag{11}$$

Where T_S is the temperature of the sample and T_E is the temperature of
the environment.

Newton's law can usually approximate the cooling of a vessel occurring by conduction or by convection in a draught, if the temperature difference is small:

$$\frac{\mathrm{d}T}{\mathrm{d}t} = K'(T_\mathrm{s} - T_\mathrm{E}) \qquad (12)$$

where K' is a constant for the system and conditions. This law also shows that, generally, if a system is perturbed at time $t = 0$ so that a temperature difference ΔT_o is reached, the system will go back to equilibrium exponentially according to

$$\Delta T_t = \Delta T_\mathrm{o} \exp\!\left(\frac{-t}{\tau}\right) \qquad (13)$$

Where τ is the time constant of the system. The use of single or multiple thermocouples in calorimetry is important. If the difference between the temperatures of the sample and reference thermocouple junctions is ΔT, then an emf (E) is produced which depends on the thermoelectric constant ε and the number of thermocouples (N). Therefore

$$E = N\varepsilon\Delta T \qquad (14)$$

The conduction of heat through the thermopiles produces a power output that changes with time, shown by the Tian equation:

$$P = \frac{k}{N\varepsilon}E + \tau\frac{\mathrm{d}E}{\mathrm{d}t} \qquad (15)$$

The exponential decay of the power with time may be integrated to measure the heat produced. Cooling in the thermopiles may be induced by passing a current in the reverse direction allowing a more efficient means of temperature control, reducing experimental time. Such principles are used in Calvet or thermoelectric heat pump calorimeters.

CALORIMETERS

As noted above, it is possible to determine enthalpy changes and other thermodynamic functions indirectly. However, this chapter will concentrate on the direct determination of enthalpy changes using non-scanning calorimeters. The construction and applications of scanning calorimeters is described in Chapter 3.

Classification of Calorimeters

Because of the profusion of calorimeter designs there is no agreed system of classification. Hemminger and Höhne[14] have suggested a method based on three criteria:

  (a) The measuring principle
  (b) The mode of operation
  (c) The principle of construction

Any calorimeter has basically two regions – the sample and the surroundings. The "sample" with a temperature T_S refers not only to the process under investigation, for example a phase change or a reaction, but also the associated containers, heaters and thermometers. The "surroundings" refers to the controlled region around the sample with a temperature T_E. The temperature control of the surroundings may be active, as in the case of a Peltier unit, or passive, in the case of a heat sink. T_E does not refer to the general laboratory conditions, which may, however, need to be controlled to minimise unwanted effects. A crucial element of calorimetry is the measurement of T_S and T_E, and their difference, ΔT, as functions of time t.

$$\Delta T = T_S - T_E \tag{16}$$

The Measuring Principle

- *Heat conduction calorimeters* operate at constant temperature. Heat liberated from a reaction is, to a good approximation, entirely diluted within a heat sink. In the case of Lavoisier's ice calorimeter, the heat of a reaction results in melting of ice, allowing the reaction to be followed. In addition, the endothermic phase change, associated with melting of ice, maintains the system at constant temperature. Modern isothermal calorimeters measure the conduction of heat as it travels between the reaction ampoule and the surroundings and they often have a very high degree of sensitivity.
- *Heat accumulation calorimeters* allow a rise in temperature of the reaction system for exothermic reactions or a decrease in temperature for endothermic reactions. A reaction is followed by measurement of a temperature change as a function of time, although modern calorimeters allow the signal to be converted into power. An adiabatic solution calorimeter is typical of this class.
- *Heat exchange calorimeters* actively exchange heat between the

sample and surroundings often during a temperature scanning experiment. The heat flow rate is determined by the temperature difference along the thermal resistance between the sample and surroundings. Heat flux DSC uses this principle.

The Mode of Operation

Three modes are important:

- *Isothermal*, where sample and surroundings are held at a constant temperature ($\Delta T = 0$, T_S = constant).
- *Isoperibol*, or constant temperature jacket, where the surroundings stay at a constant temperature while the sample temperature may alter. ($\Delta T \neq 0$, T_E = constant).
- *Adiabatic*, where, ideally, no heat exchange takes place between the sample and surroundings because they are both maintained at the same temperature, which may increase during the reaction. ($\Delta T = 0$, $T_S \neq$ constant).

The Principle of Construction

The construction may have a single measuring system, or a twin or differential measuring system. Simple solution calorimeters have a single cell, while a DSC has twin cells and operates in the scanning mode. The use of twin cells reduces the effects of internal and external noise and transient fluctuations.

Although calorimetry is intimately associated with thermodynamics, isothermal or adiabatic conditions are never exactly achieved. Allowances or corrections are made for the slight differences between theoretical and actual behaviour. Systematic errors, which may be unsuspected, can cause problems.

Calibration

A specific instrument must always be calibrated in some way. An electrical method may be available as part of the overall "package" supplied and this is always a good start. It does not, however, eliminate unsuspected systematic errors and results should always be checked by measurements of a known "standard" system that is similar to the system under investigation. This is particularly relevant for modern microcalorimeters where a signal, often of the order of a few microwatts, can be much influenced by systematic errors. The method of calibration will be noted for each type of calorimetric experiment.

INSTRUMENTATION: CALORIMETERS FOR SPECIAL PURPOSES

A very large variety of calorimetric designs have been used.[2–6,14–18] Often the design reflects the specific purpose for which the measurement is to be made. Some typical examples are listed below.

Solution Calorimeters

Solution calorimeters are usually adiabatic calorimeters. They are mainly used for the study of rapid reactions, for example, heats of solution, heat capacity of liquids, heat capacity of solids by a method of mixtures, or the enthalpy change of rapid reactions in solution. A schematic diagram is shown as Figure 3. The temperature sensor, plus a means of electrical calibration and a device for mixing reactants are all enclosed within a Dewar flask, or other adiabatic assembly.

Calibration may be performed by electrical heating or by using a reaction of known enthalpy change, such as the neutralization of a strong acid by a strong base. The whole system is allowed to equilibrate at some initial temperature T_0, and power P is supplied to the calorimeter for a time t, raising the temperature to T_1. The total energy supplied raises the temperature of the system according to:

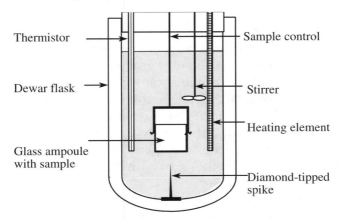

Figure 3 *Schematic of an adiabatic solution calorimeter. The assembly may be kept in a thermostated bath. Activation is done by impaling the glass ampoule onto the spike, releasing the sample into the contents of the Dewar flask*

$$Pt = C_{p, \text{total}}(T_1 - T_0) \tag{17}$$

where, $C_{p,\text{total}}$ is the heat capacity of the whole system. A measurement of the temperature change $(T_1 - T_0)$ is thus proportional to the reaction heat flow (power). The total heat capacity of the system is the sole unknown in (17) and successive experiments lead to a curve of $C_{p,\text{total}}$ as a function of temperature. $C_{p,\text{total}}$ is equal to $C_{p,o}, + m_S C_{p,S}$, where $C_{p,o}$ refers to the empty calorimeter assembly. Two sets of measurements are needed so that "empty" data can be subtracted from those for the loaded system to give the specific heat capacity $C_{p,S}$ of a sample of mass m_S. Subtraction of data for full and empty sample holders is a common requirement in calorimetry. An example may help to explain this.

Passing a current of 0.15 A through a resistor of 1000 Ω for 60 s results in a temperature rise of 3.0 °C from 22 to 25 °C in a Dewar flask containing 100 cm^3 of distilled water. The heat supplied, calculated from equation (5) is 1350 J, so that $C_{p,\text{total}}$, the average heat capacity of the system is 450 J K^{-1}. Since the heat capacity of the water is about 4.18 J K^{-1}, $C_{p,S} = 418$ J K^{-1} and $C_{p,o} = 32$ J K^{-1}.

If the temperature interval includes a phase transition, the final *apparent* sample $C_{p,S}$ may appear exceptionally large because it includes a contribution from $\Delta_{tr}H(T_{tr})$ where tr = fusion or transition at some intermediate temperature T_t as well as from low and high temperature phases (subscripts 1 and 2 respectively):

$$C_{p,S}(T_t - T_{t0}) = C_{p,1}(T_{tr} - T_0) + \Delta_{tr}H(T_{tr}) + C_{p,2}(T_t - T_{tr}) \tag{18}$$

Equation (18) shows that, by adjusting start and end temperatures and thus the total energy supplied, adiabatic calorimeters may be used to measure both C_p and $\Delta_{tr}H$.

Classical adiabatic calorimeters, with volumes of the order of tens of cm^3, are a vanishing breed. Although perhaps inevitable, this is unfortunate because the technique requires only electrical measurements and calibration using known material standards is not necessary, in principle. Many current calibrants have been characterised in this type of calorimeter. To ensure that data are always taken in near-equilibrium conditions, and apply to a definite temperature, measurements should be made under isothermal conditions, or with the lowest heating rate possible.

Adiabatic calorimeters have been used from close to 0 up to 1000 K although no single instrument can cover this entire range. Incremental operation as described gives individual "average" C_p values over a range of 5 K or less. Most C_p–T relationships approximate to linearity over such a temperature interval and corrections are rarely necessary. The best work gives values that are accurate to $\pm 0.1\%$. Enthalpies of fusion are less accurate and a realistic figure here is probably $\pm 0.5\%$. Many simple experiments involving solution reactions are described in the literature[19-21] but for the most accurate work considerable care and precautions are necessary.

Combustion Calorimeters

The enthalpy change that occurs on the combustion of a material in a reactive atmosphere, usually oxygen under pressure, is a most important method of obtaining thermochemical data, both for the stability of materials and also for the characterisation of the energy content of foods and fuels. Complete combustion of organic compounds to $CO_2(g)$ and $H_2O(l)$ is essential, and corrections may be needed for elements such as nitrogen and sulfur.

In an adiabatic flame calorimeter, the fuel, for example natural gas, or oil, is burnt completely in an atmosphere of oxygen at 1 atmosphere pressure, and the heat is transferred into a known mass of liquid, usually water and the temperature rise ΔT is measured. Provided heat losses are minimised, the enthalpy change (ΔH) per mole of fuel is given by the equation:

$$\Delta H = -C_S \frac{\Delta T}{n} \tag{19}$$

where C_S is the heat capacity of the system, obtained by calibration, and n is the moles of fuel consumed during the experiment, found from measurements of the volume or mass of fuel consumed. In flow calorimetry, a steady state is set up where a constant input of power, either by means of electrical heating or by burning a fuel supplied at a constant rate is balanced by efficient heat transfer to a heat sink.

A diagram of a flame calorimeter for combustion is shown in Figure 4. This type of calorimeter is used in several ASTM methods to determine the calorific value of gases or other fuels, for example D 1826 (1994).

As an example, consider the apparatus of Figure 4 above, where gas is supplied at 9×10^{-6} m^3 s^{-1} and water flows at 90 g s^{-1}. A temperature rise of $0.9\,^\circ$C, at steady state, was measured. Assuming the heat capacity of water is 4.18 J g^{-1} K^{-1}, the heat produced per second is:

$$dq/dt = 90 \times 4.18 \times 0.9 = 338.6 \text{ W}$$

The calorific value of the gas is 37.6 MJ m^{-3}, or, assuming ideal gas

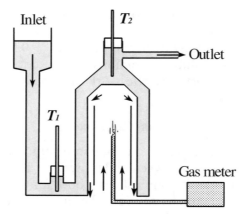

Figure 4   *Schematic of a flame calorimeter for gases. The combustion of the regulated gas flow raises the temperature of the water from* T_1 *at the inlet to* T_2 *at the outlet*

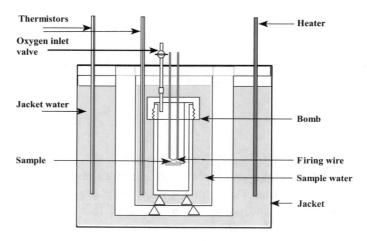

Figure 5   *Schematic of an adiabatic bomb calorimeter*

behaviour, $\Delta_c H = -843$ kJ mol^{-1}.

More widely used is the adiabatic bomb calorimeter, shown in Figure 5. The sample is placed in the crucible, in contact with an ignition wire. The crucible is placed in a strong bomb of stainless steel, which is sealed firmly and pressurised to about 25–30 atm with oxygen. The bomb is placed inside a container with a measured amount of water. This is carefully supported within a shielding container with an outer jacket. This shield contains an electrically conducting solution with a heater, or some other means of temperature control. Thermistor sensors measure the temperature of the inner container, and of the outer jacket. Having allowed the system to equilibrate until there is no difference in temperature between the sensors, the bomb is "fired" by passing a current through the ignition wire. This burns in the oxygen and ignites the sample, which combusts completely. The temperature of the outer shield is raised to match the temperature within the bomb, maintaining the system under adiabatic conditions.

The calorimeter is calibrated using a substance whose internal energy change ΔU of combustion is accurately known. Benzoic acid is often used, having

$$\Delta U = -3215 \text{ kJ mol}^{-1} = -26.35 \text{ kJ g}^{-1}.$$

Corrections for the burning of the ignition wire, and for the conversion of any nitrogen in the sample into nitric acid should be applied.

It must be noted that the bomb is a constant volume system, and therefore the calorimetry determines ΔU. Equation (20) can be used to convert ΔU into ΔH:

$$\Delta H = \Delta U + \Delta n R T \tag{20}$$

where Δn is the increase in the number of moles of gas during the combustion.

Applications of combustion calorimetry include the determination of the calorific value of fuels, and the energy content of foods, as well as obtaining primary thermochemical data, such as the resonance energy of aromatic compounds.[22,23] Bomb calorimetry is used in several ASTM methods, for example D 2382 (1988) and D 4809 (1995).

A further example of a combustion calorimeter, although of a very specific nature, is the cone calorimeter.[24] This operates using the oxygen consumption principle, which states that for most combustible materials there is a unique relation between the heat released during a combustion reaction and the amount of oxygen consumed from the atmosphere:

$$\Delta H = -13.1 \text{ MJ (kg O}_2)^{-1}$$

Thus, using this relationship, measurement of the concentration of oxygen in the combustion product stream, together with the flow rate, provides a measure of the rate of heat release (RHR). The equipment may be applied to measure the flammability characteristics of materials such as plastics and fabrics by heating a sample ($10 \times 10 \times 1$ cm approximately) using an electric heater to provide controlled irradiation. An electric spark may provide ignition and, in addition to the measurement of heat, the mass loss of the sample, the intensity of smoke produced and the concentrations of product gases can all be measured during the course of the combustion. The heats of combustion and smoke parameters compare well with those determined by other methods and the method is now recognised as ASTM E 1354 (1994).

Reaction Hazard Calorimeters

The avoidance of hazards is of prime importance on industrial sites and in laboratory testing. The determination of the conditions under which rapid exothermic reactions, explosions and unwanted side reactions can occur is vital, so that the process may be carried out with the least risk and the most profitable production of the required material. A small-scale laboratory reactor with facilities for stirring, heating, cooling addition of reagents, sampling or recycling of products, plus measurement of the system temperature and pressure may be employed to model chemical reactions and simulate larger scale processes. The fundamental characteristics of the process depend upon the total heat of reaction, the heat capacity of the system, and the effectiveness of heat transfer. These are measured by monitoring the batch and jacket temperatures, enabling the calculation of heat flow at any time. Especially important are the possibilities of runaway heating, of explosive build up of pressure or adverse reactions. The possibility of on-line monitoring of chemical composition by spectroscopic techniques such as Fourier-transform infrared spectrometry allows the study of the chemical nature of the processes involved and adds to the usefulness of the instrument.[25,26]

The measurement of temperature and pressure is a very important aspect of process safety. Knowing the enthalpy change $\Delta_r H$ of a reaction and the system heat capacity C_p then, under adiabatic conditions, the temperature rise $\Delta T_{\text{adiabatic}}$ can be determined by:

$$\Delta T_{\text{adiabatic}} = \frac{- \Delta_r H}{C_p} \qquad (21)$$

Increasing the cooling rate may contain a thermal runaway reaction. However, if the cooling fails, switching off the feed may stop the reaction. Sometimes even this will not counteract the runaway heating or pressure rise. Measurement in a calorimeter which is designed so that the heat capacity of the container $(mC)_{\text{cont}}$ is much lower than that of the chemical reactants, $(mC)_{\text{chem.}}$, which simulates the design in industrial plant, gives a better indication of likely runaway.[25] This is sometimes referred to as having a phi-factor close to 1.0:

$$\phi = 1 + \frac{mC_{\text{cont}}}{mC_{\text{chem}}} \qquad (22)$$

One system, which combines calorimetric measurements with monitoring of temperature and pressure, encloses the sample in an inert cell, to which pressure and temperature sensors are attached. The cell is then enclosed in a bomb-like container and heated at a slow rate, between 0.5 and 2°C min^{-1}, by the integral furnace. Any rapid rise in pressure or in temperature during the heating indicates that the sample is undergoing a degradation, which may pose a risk.[27]

Isothermal Calorimeters

Isothermal calorimeters, occasionally referred to as heat conduction or as heat flow calorimeters, maintain a reaction system close to constant temperature. Necessary for the measurement of heat conduction a temperature gradient is propagated from the reaction ampoule to or from a surrounding heat sink. Correction is made for this small imbalance of temperature. Peltier units are located within the temperature gradient of a reacting system, between the reaction ampoule and surrounding heat sink, so that the transmission of energy can be measured as shown in Figure 6. Various designs of heat sinks in heat conduction calorimeters have been used. These include a thermostatted water/oil bath, as in the case of the Thermal Activity Monitor[15] and the Calorimetry Sciences Corporation series of instruments,[16] thermostatted air box used for the LKB batch calorimeters[17] and temperature-regulated Peltier units as used in the Setaram.[18] Modern instruments have a twin ampoule configuration incorporating a reaction ampoule linked to a reference ampoule. The observed calorimetric signal is the differential of the two signals eliminating a considerable amount of external noise.

Such calorimeters are highly sensitive and have exceptional long-term base-line stability. This is illustrated in Figure 7 where a chemical reaction was run for 150 days in the calorimeter. The resulting signal shows little deviation as a consequence of external noise.

Calibration of isothermal calorimeters is made using an electrical heater. A known amount of power is supplied over a given time period to which the calorimetric signal is set. Some advantage is gained using chemical calibration and there have been several approaches for such purpose.[28,29] Briggner *et al.*[30] have given a comprehensive description of the fundamental workings of heat conduction calorimeters.

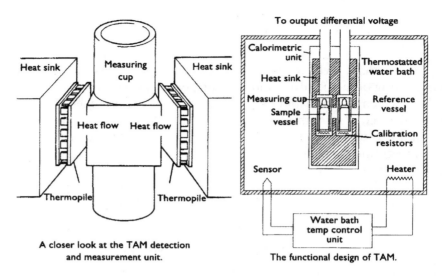

A closer look at the TAM detection
and measurement unit.

The functional design of TAM.

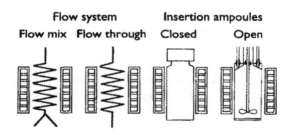

Figure 6   *Schematic of an isothermal calorimeter unit showing the position of the reaction
ampoule, Peltier units and heat sink. The instrument has different combination of
application, including flow through or ampoule insertion*
(Courtesy Thermometric Ltd, Jafalla, Sweden)

APPLICATIONS OF ISOTHERMAL CALORIMETRY

Calorimetry has found many uses within many different disciplines in the
scientific community. Over the past 8 years, the primary journals relating
to calorimetry have recorded over 24 000 references to calorimetry,
equivalent to 8 publications a day! There are active areas of research in
such diverse industries as metals, foods, agriculture, textiles, explosives,
ceramics, chemicals, biological systems, pharmaceuticals, polymers and
the nuclear industry.

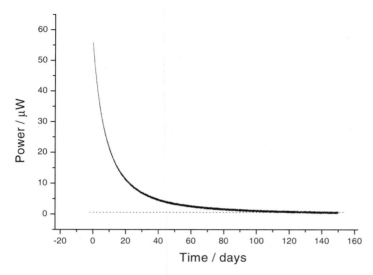

Figure 7 *Isothermal calorimeters typically have exceptional signal stability over extended time, which is useful for the study of slow reactions. Here a study of imidazole-catalysed hydrolysis of triacetin, an example of a second order reaction, was made over 150 days. A plot of power, raised to (-0.5), against time was linear, illustrating the stability of the calorimeter over the time of the study*

In the metal industry studies have been made of the phase transitions of metal alloys,[31] modelling capacitors,[32] metal hydration kinetics,[33] precipitation of solutionised aluminium,[34] and mechanisms of solid state transitions.[35]

Food applications include the shelf life of foods,[36] cooking of rice,[37] oxidation of rapeseed oil,[38] photo-calorimetric study of plants,[39] properties of amorphous sugar,[40] and metabolism of dormant fruit buds.[41]

The study of minerals systems and ceramics has given valuable information for the construction and materials industry,[42–45] and the chemical industry has also benefited from studies of reactions and physical processes.[46–51]

Calorimetric studies relating to polymers include curing of resins,[52] aggregation of PEO–PPO–PEO [PEO = poly(ethylene oxide), PPO = poly(propylene oxide)] polymers,[53] behaviour of block co-polymers.[54] Investigations of biological systems, particularly of ligand–protein interactions, have been carried out using calorimetry,[55–63] complementing DSC studies in some cases.

Pharmaceutical examples are described later, but calorimetric studies of slow reactions,[64] polymorphism[65,66] and degradation,[67–69] as well as many other applications, show the need for these techniques in the pharmaceutical laboratory.

In addition, several review articles have been published giving recent developments and current activities in isothermal calorimetry.[70–79]

CALORIMETRIC INFORMATION

The principle of calorimetry is to record the thermal events of a reaction under study. The data gathered during a calorimetric experiment has within it much useful information about the reaction, including the reaction kinetics and reaction thermodynamics, the reaction heat capacity change, the enthalpy change, the entropy change, the equilibrium constant and the reaction mechanism. The calorimetric response is a result of direct observation of the natural process involved with a reaction and imposes no interference as a result of the observation. Calorimetry is recognized as a non-invasive, non-destructive method. For a given mechanism, the reaction parameters contained within the calorimetric signal will be identical from the start to finish of a reaction. Calculation of reaction parameters from the initial calorimetric data will give the same result as the reaction parameters determined from the middle or end of the data.

The calorimetric signal measures power as a function of time and offers access to the kinetics of a reaction. Integration of the power–time signal gives the extent of a reaction in terms of enthalpy change and is a thermodynamic property of a reaction. Having both the kinetic and thermodynamic properties all that can be known about a reaction can be found. A complication is that the kinetic and thermodynamic terms are made up of several different and independent parameters. Deconvolution of these parameters can only be achieved if a mathematical model, properly describing the change taking place, is successfully applied to the data. The use of other techniques to add mechanistic information often helps in the analysis. There are, however, different strategies that can be used to discover something useful from the calorimetric data depending on the type of experiment made and the information required.

The Calorimetric Signal

Calorimeters essentially record power as a function of time by a direct measurement or by conversion from temperature changes. Both kinetic and thermodynamic information is accessible from the same calorimetric data set. The calorimetric signal, power as a function of time, is a measure of the rate of the reaction as shown in (23):

$$\frac{dq}{dt} = \frac{dx}{dt}\Delta H \tag{23}$$

where dq/dt is power in W, dx/dt is the rate of increase of the amount reacted and ΔH is the enthalpy change. The integral of power with respect to time gives the cumulative enthalpy change (q_t) at any time, so that

$$q_t = x\Delta H \tag{24}$$

where x is the quantity of material reacted at time t.

The reaction rate, defined as dx/dt, is proportional to the quantity of reactant that is available for reaction. For solution reactions, the reaction volume V should be included in the equation. Equation (25) describes a reaction where (A) reacts to form (B) in a single mechanism:

$$\frac{dx}{dt} = kV^{1-n}(A - x)^n \tag{25}$$

where dx/dt is the rate of reaction, k is the reaction rate constant, V is the reaction volume, A is the initial quantity of reactant, $A - x$ is the amount of reactant remaining at time t and n is the reaction order. Note, in some texts, A and x are expressed as concentrations, in which case V in equation (25) would be dropped. Such equations can be written in terms of power (dq/dt) by substitution of dx/dt from (23):

$$\frac{dq}{dt} = k\Delta H V^{1-n}\left(A - \frac{q}{\Delta H}\right)^n \tag{26}$$

This simplifies to

$$\frac{dq}{dt} = k\Delta H^{1-n}\,V^{1-n}(Q - q)^n \tag{27}$$

where Q is the enthalpy change for the reaction and is defined as $Q = A\Delta H$.

The integral of equation (27) provides a relationship between power and time, the calorimetric signal:

$$\frac{dq}{dt} = (kt\Delta H^{1-n}\,V^{1-n}(n-1) + Q^{1-n})^{\frac{n}{1-n}}k\Delta H^{1-n}\,V^{1-n} \tag{28}$$

It is possible to derive the parameters n, k, and ΔH from the calorimetric data and hence predict the course of a reaction from start to finish. Equation (28) is useful for simulating calorimetric data, where values of k, ΔH, A, V and n can be used to construct a model calorimetric curve and determine the course of a reaction for given reaction variables.

These types of equations can be applied to more complex reaction schemes such as sequential reactions,[80] parallel reactions[81] and solid state reactions.[82–84] Table 1 shows some of the reaction schemes where such equations can be written.

As indicated above, modern isothermal calorimeters, for example the Thermometric TAM, can detect an enthalpy change of the order of 50 nW. Such sensitivity can be useful for the study of slow degradation reactions. For example, the pharmaceutical industry would find use for an analytical technique with the capability to detect degradation of the order of 2% or less per year from an observation period of a few days. As an illustration, a 2 g drug substance with molecular weight of 500, reaction enthalpy change of 50 kJ mol^{-1}, and involving a second order reaction would have a total enthalpy change of 200 J. If the reaction had a degradation rate of 2% per year, the degradation enthalpy change would be in the order of 10.9 mJ per day. Such a reaction would give a signal of 0.127 μW, with sufficient change over a 24 h period to allow quantitative information to be obtained from the data. Willson and Beezer[13,85] have significantly advanced such analysis. A mathematical computer program that solves for n, k and ΔH from isothermal calorimetric data for reactions with a single mechanism can be obtained from the author.[86]

Reaction Mechanism

Change may come about only when a mechanism for change is available. A mechanism for change is unique to a given reaction and defines how reactants may come together to form products. The many millions of reaction schemes for chemical processes may be grouped, in general terms, within families of reaction mechanisms. Elucidation of a specific reaction mechanism gives an opportunity for prediction of the course of a reaction from start to finish. In addition, the mechanism can provide information about the propensity and rate for a reaction to take place. Having the reaction mechanism permits extrapolation of the course of reaction outside the conditions of the study.

A primary aim for calorimetric studies is to gather information about a reaction. Often this involves attributing a reaction mechanism to the calorimetric data. There are two basic approaches to this aim. The first is to gather kinetic and thermodynamic information about a reaction. These data are then fitted to a number of reaction mechanisms that, by experience, are pertinent to the reaction scheme.[87] Having decided on the mechanism that best fits the experimental data, subsequent analysis should be made to confirm the mechanism. A second approach is to use a

Table 1 *A selection of general reaction schemes showing the transformation for application towards calorimetric data*

Reaction scheme	Kinetic expression	Calorimetric expression
A → B	$\dfrac{dx}{dt} = kV^{1-n}(A-x)^n$	$\dfrac{dx}{dt} = k\Delta H.V^{1-n}\left(A - \dfrac{q}{\Delta H}\right)^n$
A + B → C	$\dfrac{dx}{dt} = kV^{1-(m+n)}(A-x)^m(B-x)^n$	$\dfrac{dq}{dt} = k\Delta H.V^{1-(m+n)}\left(A - \dfrac{q}{\Delta H}\right)^m\left(B - \dfrac{q}{\Delta H}\right)^n$
A ⇌ B	$\dfrac{dx}{dt} = k\left(\dfrac{A}{x_e}\right)(x_e - x)$	$\dfrac{dq}{dt} = k\Delta H\left(\dfrac{A}{x_e}\right)\left(x_e - \dfrac{q}{\Delta H}\right)$
Ng equation[a]	$\dfrac{dx}{dt} = kA\left(\dfrac{x}{A}\right)^m\left(1 - \dfrac{x}{A}\right)^n$	$\dfrac{dq}{dt} = k\Delta H.A\left(\dfrac{q}{A\Delta H}\right)^m\left(1 - \left(\dfrac{q}{A\Delta H}\right)\right)^n$
Autocatalytic	$\dfrac{dx}{dt} = kV^{-1}(A-x)(x_c - x)$	$\dfrac{dq}{dt} = k\Delta H.V^{-1}\left(A - \dfrac{q}{\Delta H}\right)\left(x_c - \dfrac{q}{\Delta H}\right)$
Coagulation	$\dfrac{dx}{dt} = kV^{-1}(z-x)^2$	$\dfrac{dq}{dt} = k\Delta H.V^{-1}\left(z - \dfrac{q}{\Delta H}\right)^2$

x_e is the equilibrium concentration of product. [a] The Ng equation may be used to analyse thermal decomposition and other solid state reactions. x_c is the initial amount of catalyst at the outset of the reaction. z is the initial number of particles of a specific radius prior to coagulation.

model-free method of analysis,[88] using the experimental data to calculate an activation energy. The inclusion of subsidiary information about a reaction aids choosing the most appropriate mechanistic model to fit the data. In either case, the correlation between the experimental data and the prediction of the reaction course depends on the correct choice of a mechanism that describes the calorimetric data.

Kinetic Information

The rate of reaction is extremely important where degradation rates and shelf life are required. Reaction rates are an essential measurement both from a legislative requirement and as a means of quality assurance.

Although thermodynamics can identify the potential for change, it cannot be used to predict the rate of change. As an illustration, consider the reactions listed below:

(I) $2O_3$ (g) $= 3O_2$ (g) $\Delta_r G^\ominus = -326 \text{ kJ mol}^{-1}$
(II) $2H_2$ (g) $+ O_2$ (g) $= 2H_2O$ (g) $\Delta_r G^\ominus = -486 \text{ kJ mol}^{-1}$
(III) $2Fe^{2+}$(aq) $+ H_2O_2$ (aq) $= 2Fe^{3+}$(aq) $+ 2OH^-$(aq) $\Delta_r G^\ominus = -34.4 \text{ kJ mol}^{-1}$

At 298.15 K, all the reactants shown above (I–III) have the potential to react[10] and, if left long enough, all will go to equilibrium. However, at room temperature, only reaction III proceeds at an appreciable rate and might be studied by calorimetry. To determine the rate of a reaction the kinetics must be measured. Table 2 attempts to scale reaction rates in terms of a rate constant. The rate of a reaction is dependent on the quantity of material available and on the temperature. Equation (25) gives the mathematical form of such equations where the rate constant includes the temperature dependence through the Arrhenius relationship. The ability for calorimeters to detect slow reactions is limited by the sensitivity of the instrument. The limitation for detection of fast reactions is dependent on the time constant of the instrument. For slow reactions, the Thermometric TAM can detect reaction rates of the order of 0.03% per year,[81] which is useful for monitoring slow reactions such as pharmaceutical drug degradation.

The response of the calorimeter depends on the instrumental time constant, as given in (13). In general, useful kinetic information can be gained only when the rate of a reaction is significantly slower than the time constant. However, a mathematical correction can be made for reactions that are slightly faster than the instrument time constant.[89]

The temperature dependence of a reaction rate is evident from the Arrhenius equation. As a rough rule of thumb, a reaction rate may be

Table 2 *Reaction rates scaled in terms of the rate constant. (Here, the scale refers to chemical degradation type reactions lasting from minutes to decades. For simplicity, the scale relates to a first order rate constant, and is, therefore, exponential)*

First order rate constant/s^{-1}	Half-life	Reaction rate (%)	Relative reaction rate
1×10^{-2}	1×10^{-2} s	$> 1\,s^{-1}$	} Fast
1×10^{-3}	693 s	$1\,s^{-1}$	
1×10^{-4}	1.9 h	$30\,h^{-1}$	
1×10^{-5}	19.25 h	$3.5\,h^{-1}$	} Medium
1×10^{-6}	8 days	$8\,day^{-1}$	
1×10^{-7}	11.5 weeks	$5.8\,week^{-1}$	
1×10^{-8}	2.2 years	$2.4\,month^{-1}$	
1×10^{-9}	22 years	$3\,year^{-1}$	} Slow
1×10^{-10}	222 years	$0.3\,year^{-1}$	
1×10^{-11}	2207 years	$0.03\,year^{-1}$	

doubled for each 10 °C raise in temperature.[90] The activation energy and pre-exponential factor may be found by plotting ln k against $1/T$.

Changes studied by calorimetry are either chemical reactions or physical changes. As a general observation, chemical reactions tend to be slow with a small calorimetric signal starting at a maximum, declining with time. Reaction duration tends to be of the order of days to years. Physical changes, such as crystallisation and dissolution, tend to be fast and have large Gaussian-type calorimetric signals. Reaction duration tends to be in the order of hours. Vapour sorption tends to be fast, adsorption is exclusively exothermic and desorption endothermic. Sorption events also tend to have a Gaussian-shaped calorimetric response.

EXAMPLES OF ISOTHERMAL CALORIMETRIC APPLICATIONS

Chemical Reactions

An arbitrary distinction can be made between chemical reactions and physical changes. Both sorts of change can be equally harmful to the quality of a material.

Solution Phase Reactions

Solution phase reactions are, in general, less complex than solid state reactions to study by calorimetry. Reactions are often comparatively fast

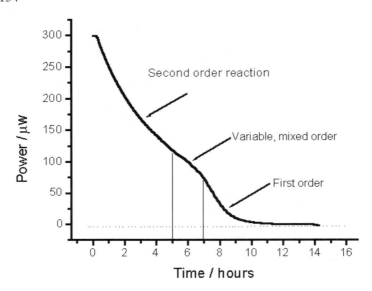

Figure 8 *Isothermal calorimetric study of the solution phase reaction of ethyl ethanoate*

with reasonably straightforward reaction mechanisms. There are many references to solution calorimetric studies in the literature.[91–94]

The hydrolysis of ethyl ethanoate[95] is often provided as an undergraduate practical analysis because of its apparent simplicity. The study of this reaction by isothermal calorimetry shows that a change in reaction mechanism occurs during the course of the reaction. A test reaction was made by mixing 2 cm^3 of a 0.011 M ethyl ethanoate solution with 0.2 cm^3 of a 1 M solution of sodium hydroxide. The resulting calorimetric signal is shown in Figure 8. The analysis of the data indicates there are three regions along the calorimetric signal, consistent with:

- an initial second order process,
- a period where the analysis of the calorimetric data gave constantly changing results as a function of time,
- a final region where the reaction became first order and the reaction went to completion.

The analysis of the data from 0–5 h gave the results in Table 3, which compare well with the literature value given.

A rationale of these observations is that the reaction is initially first order with respect to ethyl ethanoate and first order with respect to hydroxy ions, hence second order overall. As the reaction proceeds and

Table 3 *Rate data for the hydrolysis of ethyl ethanoate. Tabulated are values for rate constant, enthalpy change and reaction order calculated from calorimetric data*

Time/h	Calculated rate constant	Published rate constant[95]	Calculated enthalpy change ΔH / kJ mol^{-1}	Reaction order
0–5	$7.3 \times 10^{-2}\,\text{mol}^{-1}\,\text{dm}^3\,\text{s}^{-1}$	7.25×10^{-2}	-29.9	2
5–5.8	4.38×10^{-3}	–	–	^{a}
5.8–8	$6.53 \times 10^{-4}\,\text{s}^{-1}$	–	-28.8	1

^{a} Variable, mixed order.

the concentration of reactants declines, the order with respect to hydroxy ions decreases, and finally the reaction becomes first order in ethyl ethanoate alone.

Solid State Reactions

The majority of solid state reactions studied by calorimetry are morphological changes rather than chemical changes. This is a reflection of the complexity of chemical reactions and the difficulty of extracting meaningful information from such calorimetric data. Personal experience of physical property characterisation within the pharmaceutical industry assures me that solid state morphological stability issues are common in organic solids. About 80% of new chemical entities entering early development, in the pharmaceutical industry, have a propensity for change during the rigours of normal manufacturing processes. As a point of interest, most of the above chemical entities were not initially recognised as having a potential morphological problem using conventional spectroscopic type analytical techniques commonly used for such purposes.

The morphological diversity of compounds is clearly recognised.[96] Processing may induce morphological change due to stress. Storage conditions may also provide an opportunity for change, especially when storage extends over long periods. The consequences of poor control during processing and storage are, at times, forgotten. The mechanisms of solid state reactions are complex. Molecular restriction within a solid structure hinders chemical reaction. There are numerous examples in the literature giving interpretations of solid state reactions.[97,98] However, a general method for interpretation has yet to gain popularity. Where reaction rates are relatively slow, mathematics can be used to extrapolate from the observation period (days) into the future (years) if one wishes to determine the progress of a reaction.[85]

Compatibility Studies

Isothermal calorimetry has been used as a rapid screen for pharmaceutical materials. The main advantage is that a large number of materials, for example a number of excipients, can be tested against a drug substance in a relatively short period. There have been some publications describing quantitative methods of analysis.[99] More commonly, compatibility studies are qualitative, utilising the rapid determination of the presence of a reaction using binary drug–excipient mixtures.

Physical Changes

Physical changes tend to be relatively fast and often recorded within the observation period of a study. Physical changes are typically performed by stressing a material and following any resulting change. Such applications include crystallinity, morphological change, hygroscopicity and vapour sorption.

Crystallinity

Of all solid state transitions, the crystallisation of amorphous materials provides very abundant literature.[100–102] A useful method for analysis is to couple a heat conduction calorimeter with a vapour perfusion device. The perfusion device allows control over the reaction environment, including vapour partial pressure. Crystallisation can be induced in an amorphous sample by sequentially increasing the partial pressure of a vapour in contact with the sample, effectively lowering of the glass transition temperature. On reduction of the T_g, spontaneous re-crystallisation can occur. This is typically observed in three characteristic steps: the first is vapour sorption and is usually enhanced by the high surface area of amorphous materials. The second is a crystallisation exotherm, and the third step is endothermic desorption of vapour as a consequence of surface area reduction. Figure 9 illustrates such processes.

Morphological Change

The investigation of polymorphs by practical application of calorimetry cited in the literature employs almost every type of isothermal or isoperibolic calorimetry. The number of polymorphic forms of a crystalline material is dependent on how many different orientations the molecules can be arranged to form a crystal lattice. There is, at present, no way to predict the number of polymorphic forms a material may have.

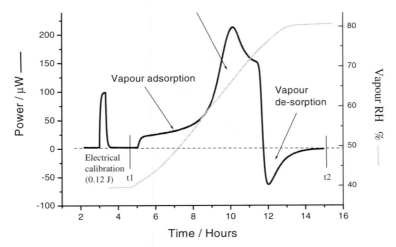

Figure 9 *The calorimetric response for vapour sorption of an amorphous drug substance with a glass transition temperature of about 95 °C. The experiment was performed in an isothermal microcalorimeter coupled with a vapour perfusion device at 25 °C. A linear RH ramp was imposed at 5% RH h⁻¹. The resulting signal shows initial moisture adsorption, recrystallisation and subsequent moisture desorption*

The more effort that is put into searching, the more forms may be found. The molecular arrangement of molecules in a lattice can have a significant effect on the physico-chemical properties of a material. A historical example dates back to the Napoleonic wars. Tin buttons on the tunics of solders became brittle and broke off when marching in the cold climates of northern Europe. Below 13.2 °C tin undergoes a polymorphic conversion from an (α) form that is strong and malleable into a (β) form that is fragile and brittle. Calorimetry can be used as a method for identification of polymorphic forms by exploitation of differences in physical properties, such as transition temperatures, solubility, vapour interaction, crystal lattice energy and so on. Figure 10 shows an example of this.

Hygroscopicity

Hygroscopicity[103] can only be properly defined if stated in conjunction with surface area. The units of hygroscopicity should be (moles of water) m^{-2} mol^{-1} material. As a general guide, a material with a surface area of 3000 m^2 mol^{-1}, typical of a micronised drug substance, will adsorb 0.5% by weight for a monolayer surface coverage. Sorption in excess of 0.5% corresponds to ever increasing degrees of hygroscopicity. Calorimetric investigations into hygroscopicity typically involve RH perfusion coupled with calorimetry. A material is subjected to increasing levels of water vapour and the mass increase followed by measuring the associated

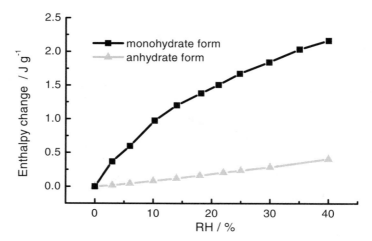

Figure 10   *Two different polymorphic forms of a drug substance showing different vapour sorption enthalpy changes. Exploitation of different thermodynamic properties of material allows the identification of a form or mixture of forms (RH = relative humidity)*

enthalpy change. Quantification of water sorption can be made using the enthalpy change for vapour condensation of 2.44 kJ g^{-1} water. An example of hygroscopicity of a pharmaceutical drug substance can be seen in Figure 11.

Vapour Sorption

Vapour sorption studies are performed extensively to characterise surface properties, such as surface energy, surface area and affinity for a given vapour.[103–105] Surface properties can be characterised by vapour sorption where the sorption data is attributed to a sorption model that conforms to basic sorption type. Brunauer *et al.*[106] have described five such types. From the adsorption model, thermodynamic information such as enthalpy change for vapour interaction and surface area can be determined. The Brunauer–Emmett–Teller (BET) type equations are typically applied to sorption data up to monolayer vapour coverage.[107,108] An extension to this equation is the Guggenheim–Andersson–DeBoer model (GAB)[109,110] and has been successfully applied to data where sorption levels exceed the formation of a monolayer. Calorimetric vapour perfusion allows good control of such experiments.[111] A plot of the enthalpy change for sorption against the vapour partial pressure allows BET type equations to be applied to the date, see Figure 12.

The mathematical form of BET equation can be transformed into an

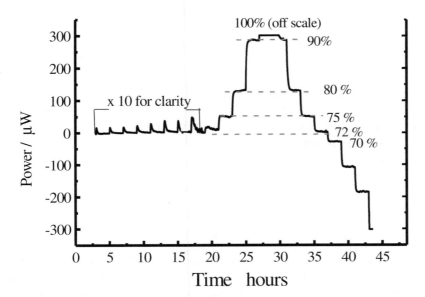

Figure 11 *Graph illustrating the application of vapour perfusion to study hygroscopicity. Here sodium chloride was used as an example. Starting at a relative humidity of 50%, the RH was stepped in increments of 2.5% to 70% RH. The resulting signal was seen to be adsorption type peaks. Above 72% RH the signal became stepped, indicating the onset of deliquescence*

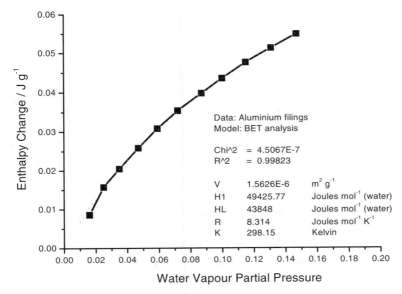

Figure 12 *Water vapour adsorption isotherm for aluminium foil. The analysis of the data using a Type II isotherm model reveals a surface area of 0.04 m² (g aluminium)⁻¹ and an enthalpy change of 49.4 kJ (mol water)⁻¹*

equation that can be applied to calorimetric data,[109]

$$q = \frac{CV_m[\Delta H_I\, x + \Delta H_I - \Delta H_I\, x^2]}{(1-x)(1-x+Cx)} \text{ where } C = \exp\left(\frac{\Delta H_I - \Delta H_L}{RT}\right) \quad (29)$$

Here V_m is the moles of water adsorbed per gram solid for the formation of a monolayer, ΔH_I is the enthalpy change for surface adsorption of water vapour, ΔH_L is the enthalpy change for water condensation, x is the partial pressure of vapour, $p/p°$, and q is the measured enthalpy change for the sorption of x.

SUMMARY

In this chapter several different types of calorimeters have been described as well as the extraordinary diverse selection of applications associated with them. It should be recognised that there is a wide variety of other calorimeters designed for a specific use. For example, the US nuclear industry makes a disposable instrument for monitoring radioactivity in nuclear power stations. Others include a calorimeter that can be placed over growing plants, in particular to measure thermo-regulation. In the realm of higher technology equipment are disposable solid state calorimeters printed on a circuit board requiring a few milligrams of sample and capable of temperature scanning at near to one million degrees per second.

The versatile nature of calorimeters, commercial and home-made, instruments allows direct access to the thermodynamic properties of materials being studied. Calorimetry is unintrusive in the way information is extracted during a study and highly versatile, measuring from nW to MW, from near absolute zero to several thousand kelvin. The sample studied can be in any phase or mixtures of phases and calorimetry can, in principle, be used to obtain all the thermodynamic and kinetic parameters relating to a reaction, and is limited only by the sensitivity of the instrument to detect a change.

Calorimeters faithfully record all the changes that occur in a sample. A temptation is to mould the resulting calorimetric signal into a preconceived idea, but although getting a result from a calorimeter is easy, to get a correct result takes a lot of time, effort and cogitation.

ACKNOWLEDGEMENTS

The author would like to thank Dr M. Richardson for his many helpful suggestions and Dr M. Phipps for his advice and expertise on isothermal microcalorimeters.

REFERENCES

1. A. L. Lavoisier and P. S. De Laplace, *Histoire de l'Academie Royale des Sciences*, 1784, **1780**, 355.
2. F. D. Rossini (ed.), *Experimental Thermochemistry I*, Wiley-Interscience, New York, 1956.
3. J. M. Sturtevant, in *Techniques of Organic Chemistry, Physical Methods*, ed. A. Weissberger, Interscience, New York, 2nd edn, 1959, Chapter XIV.
4. H. A. Skinner (ed.), *Experimental Thermochemistry II*, Wiley-Interscience, New York, 1962.
5. B. Pugh, *Fuel Calorimetry*, Butterworth, London, 1966.
6. R. B. Kemp *Handbook of Thermal Analysis and Calorimetry*, ed. M. E. Brown, Elsevier, Amsterdam, 1998, Vol. 1, Chapter 14.
7. P. W. Atkins, *Physical Chemistry*, Oxford Unversity Press, Oxford, 1978.
8. E. B. Smith, *Basic Chemical Thermodynamics*, Oxford Science, Oxford, 1990.
9. G. Price, *Thermodynamics of Chemical Processes*, Oxford University Press, Oxford, 1998.
10. R. C. Weast and D. R. Lide (eds.), *CRC Handbook of Chemistry and Physics*, CRC Press, Boca Raton, FL, 1989.
11. N. A. Lange, *Handbook of Chemistry*, Handbook Publishers, Sandusky, OH, 1956.
12. Web address for ICH guidelines. http://www.ifpma.org/ich1.html
13. A. E. Beezer, A. C. Morris, A. A. O'Neil, R. J. Willson, A. K. Hills, J. C. Mitchell and J. A. Connor, *J. Phys. Chem. B*, 2001, **105**, 1212.
14. W. Hemminger and G.W Höhne, *Calorimetry – Fundamentals and Practice*, Verlag-Chemie, Weinheim, 1984.
15. Isothermal microcalorimeter, Thermometric, Jafalla, Sweden.
16. Calorimetry Science Corporation, Provo, Utah, USA.
17. LKB-Produkter AB, Bromma, Sweden.
18. Setaram S. A. France, Caluire, France.
19. K. Grime (ed.), *Analytical Solution Calorimetry*, Wiley, New York, 1985.
20. G. P. Mathews, *Experimental Physical Chemistry*, Clarendon Press,

Oxford, 1985.

21. R. J. Sime, *Physical Chemistry: Methods, Techniques and Experiments*, Saunders College Publishing, Philadelphia, 1990.

22. C. K. Ingold, *Structure and Mechanism in Organic Chemistry*, Bell, London, 1953.

23. K. P. C. Vollhardt, *Organic Chemistry*, Freeman, New York, 1987.

24. V. Babrauskas, *J. Fire Flammabil.*, 1981, **12**, 51.

25. J. Singh, *Thermochim. Acta*, 1993, **226**, 211.

26. K. Schneider and H. Behl, *Mettler-Toledo Mag.*, 1997, (2).

27. *Technical Information Sheet #4*, Thermal Hazard Technology, Bletchley, UK.

28. R. J. Willson, A. E. Beezer, A. K. Hills and J. C. Mitchell, *Thermochim. Acta*, 1999, **325**, 125.

29. Y. H. Guan and R. B. Kemp, *Thermochim. Acta*, 2000, **349**, 163, 176.

30. L. Briggner and I. Wadso, *J. Biochem. Biophys. Methods*, 1991, **22**, 101.

31. P. Dantzer and P. Millet, *Thermochim. Acta*, 2001, **370**, 1.

32. D. V. Louzguine and A. Inoue, *Mater. Res. Bull.*, 1999, **34 (12)**, 1991.

33. B. Seguin, J. Gosse and J. P. Ferrieux, *Eur. Phys. J. Appl. Phys.*, 1999, **8**, 3, 275.

34. G. W. Smith, *Thermochim. Acta*, 1997, **291**, 59.

35. G. W. Smith, *Thermochim. Acta*, 1998, **323**, 123.

36. M. Riva, D. Fessas and A. Schiraldi, *Thermochim. Acta*, 2001, **370**, 73.

37. M. Riva, A. Schiraldi and L. Piazza, *Thermochim. Acta*, 1994, **246**, 317.

38. B. Kowalski, K. Ratusz, A. Miciula and K. Krygier, *Thermochim. Acta*, 1997, **307**, 117.

39. P. Johansson and I. Wadso, *J. Biochem. Biophys. Methods*, 1997, **35**, 103.

40. P. Lelay and G. Delmas, *Carbohydr. Polym.*, 1998, **37**, 49.

41. A. A. Gardea, E. Carvajal-Millan, J. A. Orozco, V. M. Guerrero and J. Llamas, *Thermochim. Acta*, 2000, **349**, 89.

42. O. Bonneau, C. Vernet, M. Moranville and P. C. Aitcin, *Cem. Concr. Res. (Special issue SI)*, 2000, **30**, 1861.

43. J. Salonen, V. P. Lehto and E. Laine, *J. Porous Mater.*, 2000, **7**, 335.

44. J. Salonen, V. P. Lehto and E. Laine, *J. App. Phys.*, 1999, **86**, 5888.

45. B. A. Clark and P. W. Brown, *Adv. Cem. Res.*, 2000, **12**, 137.

46. A. Stassi and A. Schiraldi, *Thermochim. Acta*, 1994, **246**, 417.

47. V. P. Lehto and E. Laine, *Pharm. Res.*, 1997, **14**, 899.

48. P. Ulbig, L. Surya, S. Schulz and J. Seippel, *J. Therm. Anal. Cal.*, 1998, **54**, 333.

49. P. Ulbig, T. Friese, S. Schulz and J. Seippel, *Thermochim. Acta*, 1998, **310**, 217.
50. C. Machado, M. D. Nascimento, M. C. Rezende and A. E. Beezer, *Thermochim. Acta*, 1999, **328**, 155.
51. Y. N. Smirnov and S. L. D'yachkova, *Russ. J. Appl. Chem.*, 2000, **73**, 870.
52. S. Gaisford, A. E. Beezer, J. C. Mitchell, P. C. Bell, F. Fakorede, J. K. Finnie and S. J. Williams, *Int. J. Pharm.*, 1998, **174**, 39.
53. A. E. Beezer, W. Loh, J. C. Mitchell, P. G. Royall, D. O. Smith, M. S. Tute, J. K. Armstrong, B. Z. Chowdhry, S. A. Leharne, D. Eagland and N. J. Crowther, *Langmuir*, 1994, **10**, 4001.
54. B. R. O'Keefe, S. R. Shenoy, D. Xie, W. T. Zhang, J. M. Muschik, M. J. Currens, I. Chaiken and M. R. Boyd, *Mol. Pharmacol.*, 2000, **58**, 982.
55. R. T. Kerns, R. M. Kini, S. Stefansson and H. J. Evans, *Arch. Biochem. Biophys.*, 1999, **369**, 107.
56. M. A. M. Hoffmann and P. J. J. M. van Mil, *J. Agric. Food Chem.*, 1999, **47**, 1898.
57. S. S. Hegde, A. R. Kumar, K. N. Ganesh, C. P. Swaminathan and M. I. Khan, *Biochem. Biophys. Acta*, 1998, **1388**, 93.
58. I. Wadso, *Thermochim. Acta*, 1995, **267**, 45.
59. R. B. Kemp, *J. Therm. Anal. Cal.*, 2000, **60**, 831.
60. R. B. Kemp, *Thermochim. Acta*, 2000, **355**, 115.
61. Y. Guan, P. M. Evans and R. B. Kemp, *Biotechnol. Bioeng.*, 1998, **58**, 464.
62. T. D. Morgon, A. E. Beezer, J. C. Mitchell and A. W. Bunch, *J. Appl. Microbiol.*, 2001, **90**, 53.
63. M. Montanari, A. E. Beezer, C. A. Montanari and D. Pilo-Valoso, *J. Med. Chem.*, 2000, **43**, 3448.
64. L. D. Hanson, *Ind. Eng. Chem. Res.*, 2000, **39**, 3541.
65. H. E. Gallis, J. C. van Miltenburg and H. A. H. Oonk, *Phys. Chem. Chem. Phys.*, 2000, **2**, 5619.
66. M. J. Starink and A. M Zahra, *J. Mater. Sci.*, 1999, **34**, 1117.
67. M. J. Koenigbauer, S. H. Brooks and C. G. Rullo, *Pharm. Res.*, 1992, 939.
68. C. Durucan and P. W Brown, *Mater. Res.*, 2000, **5**, 717.
69. T. Otsuka, S. Yoshioka, Y. Aso and T. Terao, *Chem. Pharm. Bull.*, 1994, **42**, 130.
70. I. Wadso, *J. Therm. Anal. Cal.*, 2001, **64**, 75.
71. I. Wadso, *Thermochim. Acta*, 1997, **294**, 1.
72. I. Wadso, *Chem. Soc. Rev.*, 1997, **26**, 79.
73. R. B. Kemp, *Thermochim. Acta*, 2000, **349**, XI–XII.

74. R. B. Kemp and I. Lamprecht, *Thermochim. Acta*, 2000, **348**, 1.

75. G. Buckton and P. Darcy, *Int. J. Pharm.*, 1999, **179**, 141.

76. A. E. Beezer, *Thermochim. Acta*, 2000, **349**, 1.

77. M. J. Koenigbauer, *Pharm. Res.*,1994, **11**, 777.

78. I. Wadso, *Chem. Soc. Rev.*, 1997, **26**, 79.

79. A. Beezer, J. C. Mitchell, R. M. Colgate, D. J. Scally, L. J. Twyman and R. J. Willson, *Thermochim. Acta*, 1995, **250**, 277.

80. S. Gaisford, A. K. Hills, A. E. Beezer and J. C. Mitchell, *Thermochim. Acta*, 1999, **328**, 39.

81. R. J. Willson, Isothermal Microcalorimetry, PhD Thesis, University of Kent, 1995.

82. H. Brittain, *Pharm. Technol.*, 1997, June, 100.

83. C. Li and T. B. Tang, *Thermochim. Acta*, 1999, **325**, 43.

84. M. Phipps and L. Mackin, *Pharm Sci. Technol. Today*, 2000, **3**, 9.

85. R. J. Willson, A. E. Beezer, J. C. Mitchell and W. Loh, *J. Phys. Chem.*, 1995, **99**, 7108.

86. R. J. Willson business E-mail address: RichardWillson@GSK.com

87. W. Ng, *Aust. J. Chem.*, 1975, **28**, 1169.

88. J. Opfermann, *J. Therm. Anal. Cal.*, 2000, **60**, 641.

89. P. Backman, M. Bastos, D. Hallen, P. Lonnbro and I. Wadso, *J. Biomed. Biophys. Methods*, 1994, **28**, 85.

90. F. W. Goddard and K. Hutton, *A School Chemistry for Today*, Longmans, London, 1961, p. 230.

91. R. J. Willson, A. E. Beezer and J. C. Mitchell, *Int. J. Pharm.*, 1996, **132**, 45.

92. M. J. Pikal and K. M. Dellerman, *Int. J. Pharm.*, 1989, **50**, 233.

93. G. Q. Li, S. S. Qu, C. P. Zhou and Y. Liu, *Chem. J. Chin. Univ.-Chin.*, 2000, **21**, 791.

94. E. M. Arnett, *J. Chem. Thermodyn.*, 1999, **31**, 711.

95. R. A. Fairclough and C. N. Hinshelwood, *J. Chem. Soc.*, 1937, 538.

96. A. T. Florence and D. Atwood, *Physicochemical Principles of Pharmacy*, MacMillan Press, London, 1998.

97. S. R. Byrn, *Solid State Chemistry of Drugs*, Academic Press, New York, 1982.

98. M. E. Brown, A. K. Galwey and G. G. T. Guarini, *J. Thermal. Anal.*, 1997, **49**, 1135.

99. T. Selzer, M. Radau and J. Kreuter, *Int. J. Pharm.*, 1998, **171**, 227.

100. M. J. Larsen, D. J. B. Hemming, R. G. Bergstrom, R. W. Wood and L. D. Hansen, *Int. J. Pharm.*, 1997, **154**, 103.

101. G. Buckton and P. Darcy, *Int. J. Pharm.*, 1999, **179**, 141.

102. E. A. Schmitt, D. Law and G. G. Zhang, *J. Pharm Sci.*, 1999, **88**, 291.

103. L. Campen, G. L. Amidon and G. Zografi, *J. Pharm. Sci.*, 1983, **72**,

1381.
104. G. Buckton, J. W. Dove and P. Davies, *Int. J. Pharm.*, 1999, **193**, 13.
105. J. Silvestre-Albero, C. G. de Salazar, A. Sepulveda-Escribano and F. Rodriguez-Reinoso, *Colloid Surf. A Physicochem. Eng. Asp.*, 2001, **187**, 151.
106. S. Brunauer, P. H. Emmett and E. Teller, *J. Am. Chem. Soc.*, 1938, **60**, 309.
107. S. Brunauer, P. H. Emmett and E. Teller, *Bureau of Chemistry and Solids*, George Washington University, 1938, February, Vol. 60, p. 309.
108. M. Pudipeddi, T. D. Sokoloski, S. P. Duddu and J. T. Carstensen, *J. Pharm. Sci.*, 1996, **85**, 381.
109. Z. N. Veltchev and N. D. Menkov, *Drying Technol.*, 2000, **18**, 1127.
110. N. D. Menkov and K. T. Dinkov, *J. Agric. Eng. Res.*, 1999, **74**, 261.
111. A. Bakri, *Thermometric Application Note*, TM, AB, Jafalla, Sweden, No 22021.

Chapter 6

Simultaneous Thermal Analysis Techniques

S. B. Warrington

IPTME, Loughborough University, UK

INTRODUCTION: THE RATIONALE BEHIND THE SIMULTANEOUS APPROACH

This chapter is concerned with the large family of Simultaneous Thermal Analysis (STA) techniques, in which two or more types of measurement are made at the same time on a single sample. This methodology, entailing a more complex instrument, often specially built, has been found to be essential in a variety of thermal studies, and instruments for simultaneous measurements have been constructed for more than fifty years – often as soon as it was technically possible, as the benefits of this approach were rapidly appreciated.

During the investigation of a new material it is unlikely that any single TA technique will provide all the information required to understand its behaviour. Complementary information is usually needed, which may be from another thermal technique, or other form of analysis. The challenge then is to correlate the data from the individual analyses. Regularly, throughout this book, stress is placed upon the effects of the environment of the sample on the progress of the reaction or physical transformation being studied, and therefore the form of the thermal curve. Information obtained from separate TG and DSC instruments, for instance, cannot be expected to correlate precisely when the sample is experiencing different conditions during thermal treatment. There are difficulties even in comparing data from different instruments of the same type, such as DSC instruments from different manufacturers, as is borne out by the experiences of standardisation bodies trying to develop measurement procedures that are independent of the make of instrument. This is true even for "well-behaved" systems, in which there is only one possibility for the type of event occurring on heating. When complex systems are studied, and alternative reaction schemes are possible, or the nature of the transformation is markedly influenced by the size, form, or environment of the sample, discrepancies between the sets of data are to be expected. There are also some materials, even though they may be chemically pure, that do not give precisely repeatable behaviour. The overall reaction may result in the same products each time, but the course of the transformation can differ in the details. In these cases it is advantageous to carry out the two or more measurements on the same sample. Simultaneous measurements are also valuable when dealing with inhomogeneous samples. Homogenisation, by size reduction for example, in order to obtain a more representative sample, sometimes cannot be carried out without risk of altering the system being studied.

In terms of an overall understanding of the system under study, the availability of two or more independent measurements that correlate precisely leads to a synergistic effect, so that the total value of the information is greater than merely the sum of the parts. F. and J. Paulik, in their review of their own STA studies,[1] speak of this as a "multiplying" effect.

At first sight, a simultaneous instrument may seem attractive as an alternative to buying separate units. A case has been made for this on economic grounds (initial purchase, running costs *etc.*) but the main advantages are technical, not economic. Modern simultaneous devices are capable of excellent performance, but it is still the case that they will not perform the individual measurement tasks as well as dedicated separate units. Their advantage lies in the unambiguous correlation of the two or more sets of information.

This chapter will describe the most common combinations of techniques in some detail, then look briefly at some less-widely used ones, which have nonetheless proved essential for some investigations.

SIMULTANEOUS THERMOGRAVIMETRY – DIFFERENTIAL THERMAL ANALYSIS (TG-DTA)

Instrumentation

Since TG and DTA were the two commonest techniques for many years and since DSC has taken over from DTA in recent times, and given their essentially complementary nature, it is not surprising that their combination was the first to be realised. TG is inherently quantitative, after appropriate calibration and corrections, but responds only to reactions accompanied by a mass change. DTA is capable in principle of detecting any reaction or transition that entails a change in enthalpy or heat capacity, but requires a good deal of effort before it is truly quantitative. The same is still true, though to a lesser extent, with DSC as shown in Chapter 3. Certainly for reactions involving a mass change, DTA and DSC can never be quantitative, since the material lost carries heat from the system, and the changing thermal properties of the remaining material affect the heat transfer to it. The complementary nature of the techniques can be appreciated from the processes amenable to study by each of them (Table 1).

Early attempts to couple the two techniques led to devices with separate TG and DTA sensors heated close together in the same furnace. This approach has been termed "concurrent" TA, and though an improvement over separate instruments, it is not truly STA. The main problem in mounting a DTA head in a thermobalance lies in the means of taking the DTA signal from the thermocouples without interfering with the operation of the balance. In general this is done using very fine flexible wires or ribbons.

In one of the few reviews dedicated to STA[1] the Paulik brothers claim to have been the first to build a true simultaneous TG-DTA, in 1955.[2] Over many years they developed their instrument, the "Derivatograph" to incorporate dilatometry, and evolved gas analysis, and used it to pioneer the development of controlled rate methods. By modern standards, the instrument performance was pedestrian, and used samples of

Table 1 *Processes amenable to study by TG and DTA/DSC*

Process	TG effect		DTA/DSC effect	
	Gain	Loss	Exotherm	Endotherm
Adsorption	✓		✓	
Desorption		✓		✓
Dehydration/desolvation		✓		✓
Sublimation		✓		✓
Vaporisation		✓		✓
Decomposition		✓	✓	✓
Solid–solid transition			✓	✓
Solid–gas reaction	✓	✓	✓	✓
Solid–solid reaction	Maybe	Maybe	✓	✓
Crystallisation		✓	✓	
Melting				✓
Polymerisation		Maybe	✓	
Catalytic reactions	Maybe	Maybe	✓	

up to *ca.* 1 g. The next step forward was the versatile Mettler TA-1, described by Wiedemann in 1964,[3] which had a higher sensitivity, and could be operated under vacuum or in reactive atmospheres. Stanton Redcroft (now Rheometric Scientific) and Netzsch and now several other manufacturers soon followed with instruments essentially the same as those currently available. These instruments all mount a DTA head onto the thermobalance suspension or rise rod. A simplified view of a typical sensor arrangement is shown in Figure 1.

A different approach has been taken by TA Instruments, in using two horizontal balances. The balance arms, each with a thermocouple attached to a pan carrier, hold the sample and reference adjacent to each other in the furnace. Unusually, in this instrument there is no heat flow path between the sample and reference, other than through the surrounding atmosphere, which limits the quality of the DTA data. Mettler-Toledo now offer a technique in which a form of DTA is obtained by comparing the sample temperature with a calculated reference temperature profile. Modern TG-DTA instruments are capable in general of a TG resolution around 1 μg, use samples typically from 5 to 100 mg, and can give sensitive and quantitative DTA performance when the head is of

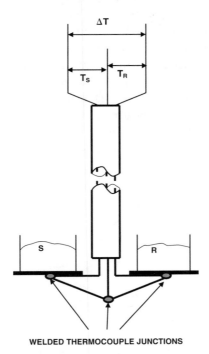

Figure 1 *Schematic diagram of DTA head mounted on a thermobalance suspension*

the appropriate type, *i.e.* there is a heat flow link between sample and reference.

Applications of TG-DTA

An example of the complementary nature of TG and DTA curves obtained simultaneously is shown in Figure 2, which represents the partial reduction of a superconducting ceramic from the "YBCO" or "123" family. The stoichiometric composition is $YBa_2Cu_3O_7$, but processing usually results in a sub-stoichiometric phase with between 6.5 and 7 oxygen atoms. Much effort went into establishing the correct tempering conditions (time, temperature, oxygen pressure) to produce the desired composition. Various approaches were tried as a means of determining the oxygen content, including reduction using a $H_2(4\%)/N_2(96\%)$ purge gas mixture. The TG curve in Figure 2 shows a multi-stage weight loss over a broad temperature range leading to a plateau at 1000 °C. The overall weight loss corresponds to the loss of *ca.* 3 oxygen atoms, but gives no clues as to the specific reduction process(es) taking place. The DTA curve shows that the first two stages are exothermic, and the second two endothermic, then, crucially, shows a sharp peak with no weight loss at 1083 °C, due to the melting of copper. The area of the peak confirms that only oxygen initially associated with copper is reduced.

Accurate temperature calibration was necessary for the example above, and this leads on to another advantage of TG-DTA instruments: the temperature calibration can be properly carried out as the sample temperature is measured directly, which is not so in conventional thermobalances in which the temperature sensor is not in contact with the sample. A TG-DTA instrument was used to provide accurate transition temperatures for the ICTAC Curie point standards for TG calibration,

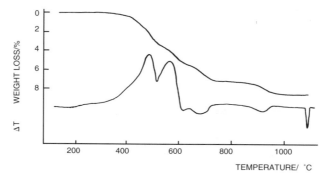

Figure 2   *TG and DTA curves for the reduction of a $YBa_2Cu_3O_7$ superconductor: 50 mg sample, 15 °C min^{-1}, 5% H_2 in N_2*

after calibrating it carefully with the melting temperatures of pure metals.[4] Precise temperature calibration of TG instruments is important not only because of the current stress on quality assurance, but also in kinetic studies. Imprecision in temperature measurement in TG work renders many kinetic studies meaningless, as an error of a few degrees (and this is easily achieved!) has a marked influence on the derived parameters, particularly when the studies are often restricted for practical reasons to a small overall temperature range.

The fact that the sample mass is continuously monitored means that DTA peak measurements can be referred back to a true sample weight, after allowing for prior lower temperature weight losses such as drying. Confidence in the quality of peak area measurements, especially at high temperatures, is increased when the TG data show that no sublimation or decomposition is taking place. Glass transition measurements can be markedly influenced by the moisture content in the case of many important polymers. With STA, the moisture content at the time of measurement is known exactly.

Figure 3 shows the results from an experiment on a coal sample. Results for combustion reactions are highly sensitive to experimental conditions, and often difficult to reproduce, making the STA approach valuable. In this case the drying of the sample is seen before *ca.* 200 °C on the TG curve, and the large exothermic effect due to oxidation can be assigned to the dry weight.

Comparison of the curves shows that a substantial proportion of the heat output occurs during a period of weight gain, information that would be difficult to obtain reliably by any other means. Interestingly, Evolved Gas Analysis (EGA) reveals also that, during this period of a net weight gain, large amounts of water and CO_2 are concurrently evolved.

DTA data can help to explain unusual features of the TG curve. Abrupt changes in the rate of weight loss can occur when the sample melts, or at

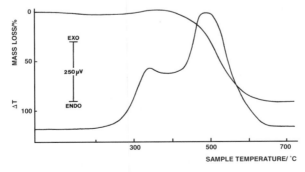

Figure 3 *TG and DTA curves for a bituminous coal: 18 mg sample, 10 °C min⁻¹, air*

least partial fusion occurs, and this may be detectable by DTA. During
the loss of material by sublimation, or solid decomposition, if the experiment extends through the melting point, the rate of weight loss will show
a slight arrest, due to the loss in surface area of the sample. The DTA
curve might indicate a fusion peak at this point. The rate of reaction in an
initially solid–state system, or the rate of a solid–gas reaction, may be
influenced by the establishment of a fluid phase. Irregular sharp features
on a TG curve due to bubbles bursting in a viscous melt are usually seen
clearly as sharp endothermic spikes on the DTA trace.

SIMULTANEOUS THERMOGRAVIMETRY – DIFFERENTIAL SCANNING CALORIMETRY (TG-DSC)

Instrumentation

Recently, the trend has been to incorporate heat-flux DSC sensors rather
than DTA sensors into the thermobalance. The greater complexity of
power-compensated sensors, and the fact that they are restricted to lower
temperatures, has precluded their use in STA. In general the DSC heat-
flux sensors are merely modified quantitative DTA plates, with the relative positions of sample and reference more precisely defined, and with a
more reproducible heat flow link between them. The implied advantage
when describing an instrument as "DSC" rather than "quantitative
DTA" is that real calorimetric measurements are possible. This is true to
some extent, and linearisation of the sensor output, and conversion of the
DTA signal into heat flow units in software or hardware reduces the effort
required, though careful calibration is again vital. One problem that may
arise, for example, is that the position of the DSC sensor may alter slightly
with different sample weights; this alters the heat transfer characteristics
and therefore the base-line of the DSC curve, and possibly the sensitivity.
Some manufacturers therefore offer the option to clamp the TG-DSC
head rigidly for best DSC performance, and thereby lose the STA ability.
The advantages of TG-DSC include all of those discussed for TG-DTA
previously, with the possibility of improved quantification of the heat
flow data.

A TG-DSC instrument capable of sub-ambient temperature operation
has been available for a number of years, and is especially valuable in the
study of systems containing moisture, or other volatiles.[5] In this case, a
DSC heat flux plate was adapted to allow it to be suspended from the
thermobalance beam, instead of the pair of thermocouples shown in
Figure 1. A completely different approach to TG-DSC is taken by
SETARAM, in their TG-DSC 111 instrument. The sample and reference

in this case are suspended in tubes surrounded by Calvet-type heat flux transducers, which do not need to contact them. The symmetrical thermobalance automatically compensates for the significant buoyancy corrections that would arise with the large samples that this instrument can accommodate.

Applications of TG-DSC

Figure 4 shows TG and DSC curves obtained from the low-temperature instrument referred to above, where the multiple solid-state phase transitions in ammonium nitrate are examined. The curves show immediately that partial sublimation, then vaporisation, of the material occurs before and after the melting point. More accurate estimates of the heats of transition and melting are thus available based on the correct sample weight. Furthermore, the literature describing the solid phase transitions is confusing, as it has been shown that traces of moisture can affect the exact course of the transitions on heating and cooling. Using STA, the moisture content is known precisely at each stage.

Setaram have described a useful application of TG-DSC in the measurement of the heat of vaporisation. The principle is illustrated in Figure 5.

The sample is sealed in a container with a small pin-hole, and, when equilibrated at the chosen temperature, loses weight at a constant rate until it is exhausted. At the same time, the DSC curve shows a deflection from the base-line equivalent to the heat of vaporisation, which can be

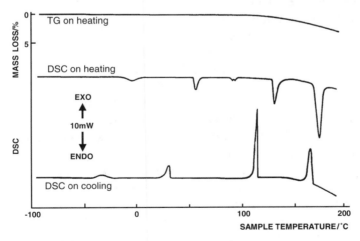

Figure 4 *TG and DSC curves for ammonium nitrate: 5 mg sample, $10\,^{\circ}C\ min^{-1}$, nitrogen*

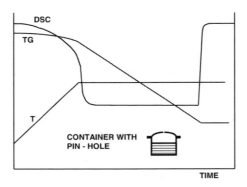

Figure 5 *Illustration of the method for measuring the heat of vaporisation by TG-DSC*

directly referred to the rate of weight loss. Calibration with materials of known vapour pressure would also allow this arrangement to derive absolute vapour presssures with good accuracy.

EVOLVED GAS ANALYSIS (EGA)

Introduction

The ICTAC definition of EGA: "a technique in which the nature and/or amount of a volatile products released by a substance are measured as a function of temperature as the substance is subjected to a controlled temperature programme" is unsatisfactory in describing the current practice of EGA as a TA technique. In many ways it is too broad, since it includes other techniques such as pyrolysis-GC, and in other ways too narrow, as it would exclude studying the changing nature of the gas stream as it is passed over the sample, as in catalytic studies. The definition does not require an evolved gas analyser to be coupled to another technique, but in practice this is nearly always the case. This section will deal with EGA as it is usually carried out, which is simultaneously, combined with another technique. The commonest combinations are those with TG and TG-DTA/DSC, where EGA assists in interpreting the chemistry of the events leading to weight losses.

There have been few reviews of the topic as a whole[6] and only one book dedicated to it.[7] The technique is, however, widely used, though the subject matter is dispersed through the literature, and not readily found. There seems to be a problem in correct choice of keywords: the official *Chemical Abstracts* term, "thermal analysis – evolved gas", is rarely used, and searches using this are disappointing.

The forerunner of EGA, Evolved Gas Detection (EGD) aimed to

determine whether a DTA peak was connected with gas loss. EGA implies either specific identification of the product and/or quantification. The value of this is easily appreciated: the main TA techniques furnish physical data, and chemical interpretation is usually obtained by inference, or analysis of the sample before and after a thermal event. This approach can be insecure, because transient intermediates may exist that cannot be isolated for analysis, or the product may transform on exposure to the atmosphere. EGA offers the opportunity of obtaining specific chemical information during the thermal experiment.

There has been a multitude of approaches to EGA, ranging from simple methods for answering a particular need to sophisticated research tools. Nearly all TA instruments use a flowing purge gas, which contains the chemical information that is sought. At the simplest level, holding a piece of wet litmus paper, or other indicator in the effluent stream from the instrument, may answer the question at hand. Beyond this, almost every type of gas detector/analyser has been linked to every type of TA equipment, and most manufacturers offer one or more methods for EGA, usually in combination with TG or TG-DTA/DSC. The approach taken in any instance depends on the information required, and there is no truly universal solution.

Instrumentation for EGA

EGA has employed almost every known type of gas detector. These may be specific, capable of responding to a single gas, or be general purpose detectors capable in principle of responding to all, or at least many gases. Specific detectors include hygrometers,[8] non-dispersive infrared cells,[9] paramagnetic oxygen detectors, chemiluminescence sensors for nitrogen oxides, fuel cell-based devices for hydrocarbons, and many others. Generally the specific detectors are capable of quantitative performance.

A useful approach that has appeared sporadically is to pass the evolved gas/purge gas mixture through an absorbing solution, and then follow the changing concentration of the product of interest by titrimetry,[10,11] ion-selective electrodes,[12] conductimetry,[13] colorimetry,[14] *etc*. These methods are relatively easy to set up, inexpensive, and capable of excellent quantitative performance. In general, linking to the TA unit will be a simple matter, though bubbling of the effluent gas through the absorbing solution may cause disturbances to a thermobalance.

Gas chromatography (GC) has been used, when the main advantage is its ability to separate mixtures. Conclusive identification of the components would be possible if a mass spectrometer (MS) was connected to the GC detector.[15] GC is necessarily a batch technique: a portion of the gas products is collected over a chosen time span, and the ability to build up a profile of the evolution of a given species is limited by how many samples can be taken in the course of the experiment. Connecting a GC to a thermobalance requires a carefully designed interface if the TG data is to remain unaffected. A simple and convenient technique has recently been described[16] in which organic vapours are trapped in an adsorbent tube for subsequent "off-line" analysis by desorption-GC-MS.

Recently, the trend has been to avoid the "home-made" approach represented by many of the above combinations, and use a system based on multi-gas detectors, primarily Fourier-transform infrared spectrometry (FTIR) or quadrupole MS. Most manufacturers can supply a thermobalance or STA unit fitted with either detector. These two techniques currently account for nearly all EGA applications in the past few years. This is possibly a reflection of the fact that off-the-shelf instruments with high specifications are quickly becoming cheaper and easier to use. Both FTIR and MS are capable of building up continuous profiles of the evolution of several species, by taking rapid scans of the IR or mass spectrum of the effluent gas repeatedly, typically every few seconds, during the experiment, then constructing EGA curves for chosen species on the basis of an IR band, or mass spectral peak, characteristic of that species. FTIR does not respond to non-polar molecules, and gases such as nitrogen and oxygen are therefore not registered. MS can detect all gases, and is considerably more sensitive. Difficulties arise with both techniques when the product mixture is complex, as the superimposed FTIR or mass spectral information for several species may be awkward, or impossible, to interpret. In that case, the solution is to use GC-FTIR or GC-MS, to separate the components before identification. On the basis of the information from the GC-based experiments, suitable characteristic absorption bands, or mass peaks, can then be chosen as representative of the species of interest for continuous monitoring, to give more detailed EGA profiles.

The method of linking the TA unit to the gas detector requires careful design to achieve good performance. Ideally the quality of the information from the thermobalance, for example, should not be compromised. This implies that the pressure inside it should not be changed significantly by the connection to the EGA unit. The aim then is to transfer evolved gases from the vicinity of the sample to the detector as quickly as possible, so that the EGA data is effectively simultaneous with, *e.g.*, the TG curve. Although transfer delays can be allowed for, apparatus with a long transfer time will result in smearing of the EGA curve due to diffusion broadening. A long residence time in the transfer system might also lead to unwanted secondary reactions. The transfer path should be inert with respect to the products of interest; certain species can degrade in contact with metal tubing, for instance, especially when it is heated, as is normally the case to avoid condensation of heavier products. Pipework of vitreous silica is the usual choice. Joining to an FTIR is relatively simple. The gas cells used create only a slight resistance to the effluent gases from the TA unit, and a short length of small-bore tubing is used to carry all the gas to the cell. Joining to a MS is more difficult, as these operate under vacuum, typically *ca.* 10^{-6} mbar. To maintain this pressure, only $1-2$ cm^3 of the effluent gas is passed to the ion source. Some form of split is therefore needed to remove the bulk of the purge gas/product mixture. The commonest solution is to use a capillary-and-bypass arrangement, as shown in Figure 6. Other approaches are a "skimmer" coupling,[17] or a jet separator,[18] which has to be used with a helium purge gas, but gives a dramatic increase in sensitivity.

Both FTIR and MS require a powerful data system which, in addition to controlling the equipment, and displaying data, may also have libraries of standard spectra to assist identification. A complete integration of the TA and EGA software is unfortunately rarely achieved. The literature up to 1997 on the techniques and applications of TG-IR has been summarised by Materazzi[19] and excellent reviews of the use of MS for EGA are available.[20-22]

Applications of EGA

A good example of what can be achieved by combining "off-the-shelf" equipment is the DTA-EGA apparatus used by Morgan for a number of years. A DTA with a large bore (37 mm) tube furnace was connected to a series of non-dispersive infrared (NDIR) detectors, specific to water, CO, CO_2, SO_2 and NH_3. A high (300 ml/min^{-1}) purge gas rate gave a good

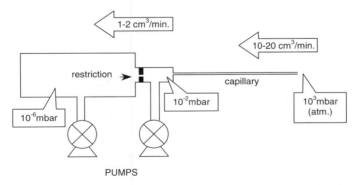

Figure 6 *Schematic diagram of a capillary-and-bypass inlet for MS sampling from atmospheric pressure*

response time, while the detectors had adequate sensitivity to cope with the dilution of the products in such conditions.[23] The apparatus was used in the assessment of the graphite content of a schist, from Burma, that also contained calcite.[24] Figure 7 shows the DTA curve recorded in oxidising conditions, and the EGA curves for CO_2 in oxidising and inert conditions.

The large exothermic peak shows a superimposed endothermic peak due to calcite dissociation. Subtracting the EGA peak recorded in nitrogen, arising from the calcite dissociation, from that in oxidising gas gives the amount of CO_2 coming only from the graphite oxidation, which then allows a quantitative estimate of the graphite content to be made.

A TG-FTIR combination has been used to study the degradation of recyclable car body underseal materials containing PVC.[25] In this case the small TG furnace, FTIR gas cell and transfer line, each of a few cm³ in volume, needed only *ca.* 20 cm³ min⁻¹ of purge gas, which resulted in high sensitivity and speedy response. A stacked plot of partial spectra collected at intervals during heating is shown in Figure 8.

The spectra show the rotational lines due to HCl released on heating. Strongly polar molecules like HCl are ideal for study by IR, as their absorbance is high, and this experiment showed clearly that the gas was released at temperatures as low as *ca.* 150 °C. The instrument was used to estimate the PVC content of the materials by comparing the EGA peak areas with those for pure PVC.

A TG-DTA-MS instrument was developed[18,26] by the author and used in a wide range of projects, covering most classes of material. A simple capillary coupling was shown to operate with all common purge gases, and up to 1500 °C. A special feature of the interface is the option to select a jet separator, which is mounted in parallel to the commoner

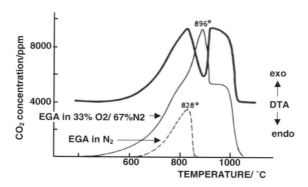

Figure 7   *DTA and EGA curves for a graphite–calcite schist: 100 mg sample, 10 °C min⁻¹, N₂ (67%) O₂ (33%)*

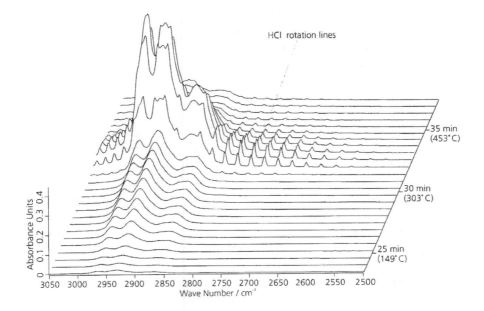

Figure 8 *Changes in the FTIR spectrum of the evolved gases with temperature during the heating of a PVC-containing material*

bypass arrangement, in a temperature-controlled enclosure. With this and the capillary heated to around 180 °C, nearly all products are transferred efficiently through the 0.3 mm ID silica connecting tube. The interface can isolate the MS from the atmosphere, in contrast to some systems that sample all the time. This has advantages in being able to bake out the analyser to reduce background signals due to contamination, and in carrying out MS diagnosis. Over time, it has become clear that the best results are obtained only with a carefully designed all-metal plumbing system. All polymeric tubing is permeable to water, oxygen, or both, which are detectable by the MS. A great virtue of EGA is its ability to detect leaks in the gas path, and monitor purging procedures which are essential to obtain a truly inert atmosphere. It is revealing to find what effort has to be expended to reduce oxygen levels in standard TA equipment to a level where they react to only an insignificant degree with sensitive materials at high temperature.

This instrument was used in a study of a pyrotechnic initiator composition,[27] based on fine (1.7 μm) zirconium powder, potassium perchlorate, and a small amount of nitrocellulose as a binder. Mixtures were studied in helium, (a) to take advantage of the higher sensitivity of the jet separator, (b) to reduce the likelihood of ignition of these sensitive mixtures by using a purge gas of high thermal conductivity and (c) to use the efficient purging properties of helium in sweeping traces of oxygen from the internal spaces of the equipment. TG, DTA and EGA curves for one composition are shown in Figure 9. The curves represent a complex sequence of events.

First the nitrocellulose decomposes around 200°C, leaving a carbon rich residue of only *ca.* 10% of the weight of nitrocellulose. Following the sharp orthorhombic–cubic phase transition in potassium perchlorate at 300°C, an exothermic reaction takes place between the zirconium and perchlorate while, at the same time, the carbon residue is oxidised by the perchlorate, as shown by the CO_2 curve. The persistence of CO_2 evolution up to *ca.* 600°C is a real and repeatable phenomenon, and remains unexplained. Oxygen evolution from excess perchlorate starts around 400°C, where it is more sensitively detected by the MS than the TG curve. The O_2 curve dips temporarily during the fusion of first a eutectic mixture of the perchlorate and its chloride reaction product, and then the remaining perchlorate, as shown by the DTA peaks in this region. Excess perchlorate then decomposes exothermically. The sensitivity of the EGA measurements can be judged from the CO_2 peak around 400°C, which

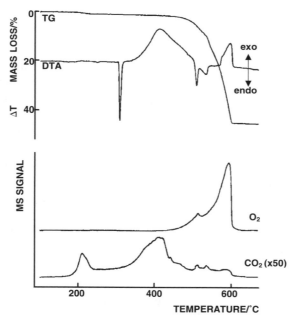

Figure 9   *TG, DTA and MS curves for oxygen and carbon dioxide, for a mixture of 10% zirconium, 89% potassium perchlorate and 1% nitrocellulose: 10 mg sample, 10°C min^{-1}, helium*

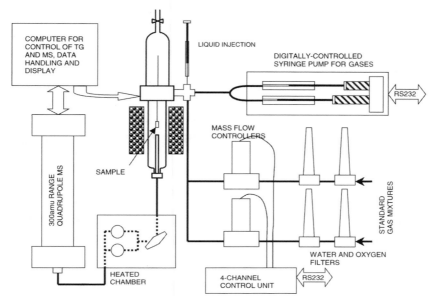

Figure 10 *TG-MS equipment showing various means of calibration of the MS sensitivity for quantitative EGA studies*

represents *ca.* 30 μg of the gas. The detailed interpretation of the course of events undergone by this mixture was possible only by using the additional information given by EGA.

The same MS instrument was also linked to a TA Instruments model 951 thermobalance, which is ideal for EGA work, but is sadly no longer available.[28] This combination was used to investigate the problem of calibration of a TG-MS for quantitative EGA. A schematic diagram of the layout is shown in Figure 10.

Calibrations could be made on the basis of the decomposition of solid standards, injections of volatile liquids, standard gas mixtures, injections of pulses of gas, or steady injection rates of gases into the purge gas stream. With proper normalisation procedures, all methods gave equivalent results.[29] A good linear relationship between EGA peak area and amount of product was found for between a few and a few hundred microgrammes of gas when using a jet separator.

LESS COMMON TECHNIQUES

Optical Techniques

The value of using an extremely sophisticated sensor (the human eye) in the study of heated materials has been appreciated for some time. In the context of TA, the monitoring of the optical properties during a thermal program can be of immense assistance. The term thermoptometry is used to cover observations, or measurements, of an optical property as a function of temperature. A useful review of the field is in the textbook by Haines.[30] Samples can be viewed by reflectance, or by transmission, and the light intensity in each case can be measured using a photocell, thereby making the technique into a measurement instead of an observation. Temperature-controlled stages for samples being viewed through a microscope (Hot-stage microscopy) have been available for many years, and modern versions are available from, *e.g.*, Mettler and Linkam. The observations are often helpful in the interpretation of other thermal data, in revealing the nature of the reaction or transition taking place. With unknown materials, a preliminary screening can avoid damage to TG or DSC equipment by showing whether the sample melts, bubbles, creeps, shrinks, swells, decrepitates, or damages the crucible. Solid–solid transitions may be detectable by a colour change, or movement due to a change in volume. Microscopic examination of reaction interfaces may suggest an appropriate choice of reaction model. Several systems have been described for combining DSC or DTA with microscopy.[31,32]

A more unusual combination is that of TG with the ability to view the sample, shown in Figure 11, which was used to study the pyrolysis and combustion of coal particles.[33]

The video camera was able to produce 640×480 pixel images of an area *ca.* 2×2 mm, with the help of a microscope with an extra-long

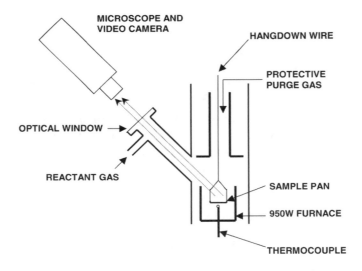

Figure 11 *Equipment for simultaneous TG and microscopic observation*

working length. Partial fusion, swelling, and changes in structure could be related to the progress of the TG curve, while particles of around 0.5 mm were heated upon on a platinum mesh support. These workers made a special furnace, for mounting in a Perkin-Elmer thermobalance, that allowed heating rates of up to $100\,^{\circ}C\,s^{-1}$, similar to those experienced by pulverised coal in power plants.

Spectroscopic Techniques

Microscope hot-stages designed for transmitted light work, and special heating accessories, have been used in conjunction with an FTIR spectrophotometer to study physical and chemical changes in a heated sample. By using a hot-stage with a DSC cell incorporated, simultaneous DSC-FTIR can be achieved.[34] The range of experiments possible in this case was restricted by the need to use thin films of sample, which resulted in poor DSC curves. More flexibility was obtained by using an IR microscope accessory to collect reflectance spectra from a sample in a DSC cell.[35] This equipment was used to follow the curing reaction in an amine-cured epoxy material, and to study structural changes in PET through its glass transition and melting regions.

X-Ray Techniques

Whereas EGA is useful in obtaining chemical information in cases where gases are lost during a reaction, it gives no direct data on the condition of the solid phases. The technique of X-ray Diffraction (XRD) has been used extensively to identify reaction intermediates and products, and can be applied to the small samples generally used in thermal studies. Extracting samples from, *e.g.*, a thermobalance for analysis might give misleading results if the samples are susceptible to reaction with atmospheric gases, primarily oxygen, water and carbon dioxide, or the solid was liable to undergo a phase transition. The ability to apply XRD simultaneously with other methods has therefore been exploited in several instances.

XRD was coupled with TG[36] by building a top-loading thermobalance onto the diffractometer in such a way as to place the sample layer in the X-ray beam. In cases where the heating rate would be limited by the scanning rate of the diffractometer, it is possible to scan only small angular regions characteristic of the products of interest, with a consequent saving of time. This type of apparatus was used to investigate the complex series of overlapping stages in the reduction, by hydrogen, of tungstic oxide, which results in a fairly featureless TG curve, but was shown by the XRD spectra to proceed *via* the production of many phases in the tungsten–oxygen system, and two modifications of tungsten.

XRD has also been coupled with DSC. An instrument based on a conventional X-ray source was described by Newman *et al.*[37] that was used to study catalyst regeneration. This instrument incorporated a mass spectrometer for simultaneous EGA. A series of XRD spectra taken during the reduction of the spent catalysts is shown in Figure 12, and reveals the intermediate product copper(I) oxide, before final reduction to copper itself.

A difficulty with the DSC–XRD coupling is that the XRD requires slow temperature scans to obtain sufficient sensitivity, but this results in lower DSC sensitivity. The problem has been solved by several workers through the use of a high-intensity synchrotron X-ray source. Caffrey, for example, used such a combination to study biological liquid crystals.[38] The need for higher intensity X-rays is greater when carrying out small-angle and wide-angle scattering experiments (SAXS and WAXS), and again synchrotron radiation has been used, as typified by the work of Ryan and co-workers on polymer structure determination.[39] Another unusual simultaneous coupling was that of Temperature-programmed reduction with XRD and Mössbauer spectroscopy, which was applied to Fe–Mo–O catalysts.[40]

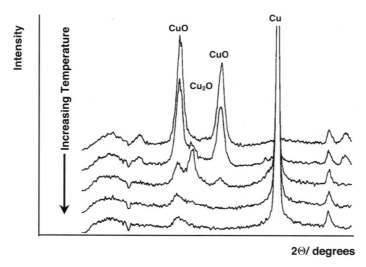

Figure 12   *XRD spectra recorded at increasing temperatures for a supported copper catalyst*

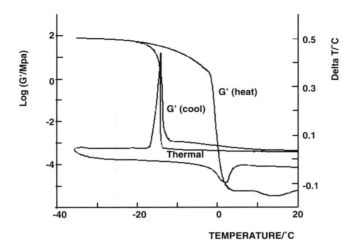

Figure 13 *Simultaneous DTA and rheology curves for a low-fat margarine*

DTA–Rheometry

Rheometry is not usually classified as a thermal analysis technique, but when data are collected as a function of temperature it must be considered as such. Correlation of temperature-resolved rheological measurements with a separate DSC experiment would present problems due to the very different sample sizes and environments employed in each technique. A recently described device[41] combines a dynamic stress rheometer with a simple DTA sensor, both heated and cooled by a Peltier element. An example of the information available is shown in Figure 13, where both the DTA and modulus are shown as a margarine sample is cooled, then heated. The rheological properties of the material are not recovered after cooling, suggesting that the structure of the emulsion has been altered irreversibly. The melting and crystallisation are revealed clearly on the thermal signal.

MICRO-THERMAL ANALYSIS (μ-TA)

The recent technique of micro-thermal analysis (μ-TA), which now has a variety of measurement modes, is included here because usually two or more measurements are made simultaneously. Micro-TA combines the imaging capabilities of atomic force microscopy (AFM) with a form of localised thermal analysis, and is able to measure thermal transitions on an area of a few microns. A good introduction to the whole family of these methods is available on the internet, from which application studies can be downloaded.[42] In brief, the method entails using a very fine platinum wire loop as the AFM tip, which acts as a heater, and a resistance thermometer. This is scanned over the sample surface to obtain the AFM topography image and, by measuring the power required to keep the tip at a fixed temperature, an image based on variations in thermal conductivity of the region below the tip. From the scanned area, points can be selected for localised thermal analysis, when the tip is applied to the surface with a constant small force, and the tip temperature is raised linearly. Thermal transitions are detected by measuring the power applied to the tip, and its position, which shows expansion or indentation during the event, or a change in the rate of these processes.

A simple example of its use is shown in Figure 14. Here a paracetamol tablet is scanned over an area of $60 \times 60\,\mu m$ without any prior sample preparation. The topographic image shows the surface morphology, but does not reveal the constituents of the tablet. The thermal image shows a sharp contrast between regions of different thermal conductivity. Localised TA shows a sharp melting peak on both the power and indentation signals, proving the presence of paracetamol. The other region gives no response up to $250\,°C$, and is therefore probably a filler particle.[43]

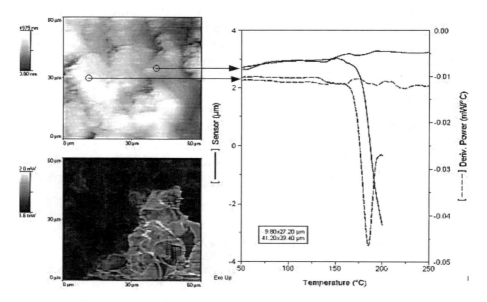

Figure 14   *Topographic (top left) and thermal conductivity (bottom left) images of the surface of a paracetamol tablet. The right-hand plot shows the results of localised microthermal analysis on the two phases revealed in the conductivity image*

The technique has been extended to give direct chemical information. The heated tip can be used to pyrolyse a small area of interest by rapidly heating to *ca.* 800 °C. The resulting plume of gas is drawn into a small sampling tube containing *ca.* 0.1 ml of Tenax and Porapak, where organic volatiles are adsorbed. The tube is then transferred to an automatic thermal desorption unit, from which the products are passed to a GC-MS system for separation and identification.[44] Applications have included the identification of polymers in multi-layer films, and the analysis of layers of emulsion paint. A recent addition to the capabilities of micro-TA is localised FTIR analysis. The area selected for examination is irradiated with an intense focused IR beam. The tip in this case acts as a bolometer, and records the interferogram, which is then deconvoluted using the FTIR software to produce an IR spectrum of a small area. The entire field of μ-TA is covered in a comprehensive recent review by Pollock and Hammiche.[45]

SUMMARY

This chapter has described the commonest simultaneous combinations, and some less common, but the coverage is by no means exhaustive, as the number of possible permutations is great. Investigators have used extraordinary skill and ingenuity in constructing special equipment to solve specific problems, and usually with much success. Most practitioners will meet simultaneous methods in the form of commercially available instruments, whose future is assured due to their value in interpreting more complicated systems. As a research tool in the thermal analysis laboratory, TG-DTA or TG-DSC are considered by the author to be invaluable. On a more basic level, their use in method development, and for multi-parameter quality control, are additional advantages. It is to be hoped that the examples presented here will stimulate readers to consider the possibility of specially-constructed apparatus, and not rely wholly on the manufacturers who are not generally in business to make "one-off" instruments.

REFERENCES

1. F. Paulik and J. Paulik, in *Comprehensive Analytical Chemistry*, ed. G. Svehla, Elsevier, Amsterdam, 1981, **XII(A)**.
2. F. Paulik and J. Paulik, *Analyst*, 1978, **103**, 417.
3. H. G. Wiedemann, *Chem. Ing.-Tech.*, 1964, **36**, 1105.
4. E. L. Charsley, S. St. J. Warne and S. B. Warrington, *Thermochim. Acta*, 1987, **114**, 53.

5. E. L. Charsley *et al.*, in *Thermal Analysis*, ed. H.G. Wiedemann, Birkhauser, Basel, 1980, Vol. 1, p. 237.

6. S. B. Warrington, in *Thermal Analysis – Techniques and Applications*, ed. E. L. Charsley and S. B. Warrington, Special Publication 117, Royal Society of Chemistry, Cambridge, 1992.

7. W. Lodding (ed.), *Gas Effluent Analysis*, Marcel Dekker, New York, 1967.

8. S. B. Warrington and P. A. Barnes, in *Proceedings 2nd ESTA Conference*, ed. D. Dollimore, Heyden, London, 1981.

9. D. J. Morgan, *J.Thermal Anal.*, 1977, **12**, 245.

10. H. U. Hoppler, *Labor Praxis*, 1991, **15**, 763.

11. J. Paulik and F. Paulik, *Thermochim. Acta*, 1972, **4**, 189.

12. T. Fennell *et al.*, in *Thermal Analysis*, ed. H. G. Wiedemann, Birkhauser, Basel, 1971, Vol. 1, p. 245.

13. S. Brinkworth *et al.*, in *Proceedings 2nd ESTA Conference*, ed. D. Dollimore, Heyden, London, 1981.

14. J. Chiu, *Thermochim. Acta*, 1986, **101**, 231.

15. P. Barnes *et al.*, *J. Thermal Anal.*, 1982, **25**, 299.

16. T. Lever *et al.*, *Proc. 28th NATAS Conf.*, Oct. 2000, Savannah, Ga, 2000, 720.

17. E. Kaisersberger *et al.*, *Thermochim. Acta*, 1997, **295**, 73.

18. E. L. Charsley, N. J. Manning and S. B. Warrington, *Thermochim. Acta*, 1987, **114**, 47.

19. S. Materazzi, *Appl. Spectrosc. Rev.*, 1997, **32**, 385.

20. O. P. Korobeinochev, *Russ. Chem. Rev.*, Dec. 1987, 957.

21. D. Dollimore *et al.*, *Thermochim. Acta*, 1984, **75**, 59.

22. K. G. H. Raemaekers and J. C. J. Bart, *Thermochim. Acta*, 1997, **295**, 1.

23. A. E. Milodowski and D. J. Morgan, *Nature*, 1980, **286**, 248.

24. D. J. Morgan, *J. Thermal Anal.*, 1977, **12**, 245.

25. E. Post *et al.*, *Thermochim. Acta*, 1995, **263**, 1.

26. E. L. Charsley, C. Walker and S. B. Warrington, *J. Thermal Anal.*, 1993, **40**, 983.

27. B. Berger, E. L. Charsley and S. B. Warrington, *Propellants, Explosives Pyrotechnics*, 1995, **20**, 266.

28. E. L. Charsley, S. B. Warrington, A. R. McGhie and G. K. Jones, *Am. Lab.*, 1990, **22**, 21.

29. S. B. Warrington, unpublished data.

30. P. J. Haines, *Thermal Methods of Analysis – Principles, Applications and Problems*, Blackie, Glasgow, 1995.

31. W. Perron *et al.*, in *Thermal Analysis*, ed. H. G. Wiedemann, Birkhauser, Basel, 1980, Vol. 1.

32. B. Forslund, *Chem. Scr.*, 1984, **24**, 107.
33. A. N. Matzakos *et al.*, *Rev. Sci. Instrum.*, 1993, **64**, 1541.
34. F. M. Mirabella, *J. Appl. Spectrosc.*, 1986, **40**, 417.
35. D. J. Johnson *et al.*, *Thermochim. Acta*, 1992, **195**, 5.
36. N. Gerard, *J. Phys. E*, 1974, **7**, 509.
37. R. A. Newman *et al.*, *Adv. X-Ray Anal.*, 1987, **30**, 493.
38. M. Caffrey, *Trends Anal. Chem.*, 1991, **10**, 156.
39. A. J. Ryan *et al.*, *ACS Symp. Ser.*, 1994, **581**, 162.
40. H. Zhang *et al.*, *J. Solid-State Chem.*, 1995, **117**, 127.
41. *Application Note*, Rheometric Scientific, Piscataway, NJ.
42. www.anasys.co.uk
43. D. M. Price *et al.*, *Int. J. Pharm.*, 1999, **192**, 85.
44. D. M. Price *et al.*, *Proc. 28th NATAS Conf.*, Orlando, FL, 705.
45. H. M. Pollock and A. Hammiche, *J. Phys. D: Appl. Phys.*, 2001, **34**, R23.

Appendices

APPENDIX 1. A   SYMBOLS FOR PHYSICAL QUANTITIES AND UNITS

Quantity	Symbol	Unit	Abbreviation
Basic physical quantities			
Length	l	metre	m
Mass	m	kilogram	kg
Time	t	second	s
Electric current	I	ampere	A
Temperature	T	kelvin	K
Amount of substance	n	mole	mol
Luminous intensity	I_v	candela	cd
Derived SI units			
Energy	E	joule	$J = kg\,m^2\,s^{-2}$
Power	P	watt	$W = J\,s^{-1}$
Force	F	Newton	$N = J\,m^{-1}$
Pressure	p	pascal	$Pa = N\,m^{-2}$
Frequency	v	hertz	$Hz = s^{-1}$
Electric potential difference	V	volt	$V = J\,A^{-1}\,s^{-1}$
Heat	q	joule	J
Heat capacity			
at constant pressure	C_p		$J\,K^{-1}$
at constant volume	C_V		$J\,K^{-1}$
Internal energy	U		J
Enthalpy	H		J
Free energy	G		J
Entropy	S		$J\,K^{-1}$

Other symbols are defined as they apply to each technique.

Nomenclature for Thermal Analysis and Calorimetry

The nomenclature for thermal analysis and calorimetry is presently under review, so that recent developments and calorimetric methods may be included. The recommendations approved by the ICTA Council and by IUPAC are reported by Dr Robert C. Mackenzie in *Treatise on Analytical Chemistry*, Part 1, Vol. 12, Section J, ed. P. J. Elving, Wiley, New York, 2nd edn., 1983. A provisional version was also given in *For Better Thermal Analysis and Calorimetry*, ed. J. O. Hill, ICTA, Edition III, 1991. However, this has not been approved by ICTA Council. A further revision is in preparation.

The ASTM have adopted a Standard Terminology Relating to Thermal Analysis (ASTM E 473-99) which is similar in many respects to the ICTAC definitions.

APPENDIX 1.B   SOURCES OF INFORMATION

1.B.1   Journals

While scientific papers containing experimental work using calorimetry and thermal analysis techniques are published in many major journals, there are some which are chiefly devoted to this field:

Journal of Thermal Analysis and Calorimetry
Published by	Kluwer Academic Publishers,
	3300 AH Dordrecht, The Netherlands
Web:	http://www.wkap.nl/journals/thermal analysis
Manuscripts to:	Lexica Ltd, Production Department for JTAC,
	H-1072 Budapest,
	Nagydiofa u.5.IV/1, Hungary
	E-mail: lexica@mail.datanet.hu

Thermochimica Acta
Published by	Elsevier Science, Journal Department,
	P.O. Box 330, 1000 AH Amsterdam,
	The Netherlands
Web:	http://www.elsevier.com/locate/tca
Manuscripts to:	Professor D. C. M. Craig, School of Pharmacy,
	The Queens University of Belfast,
	97 Lisburn Rd,
	Belfast, BT9 7BL

Journal of Chemical Thermodynamics
Published by   Academic Press, New York
Web: http://www.academicpress/jct

Netsu Sokutei (in Japanese, English summaries)
Published by   The Japan Society of Calorimetry and Thermal
                      Analysis,
                      Miyasawa Bldg 601, 1-6-7 Iwamoto-cho,
                      Chiyoda-ku
                      Tokyo 101-0032, Japan
Web: http://wwwsoc.nacsis.ac.jp/jsta/index.html

Occasional review articles are published in *Review of Scientific Instruments*, *Journal of Physics, E*, and the *International Journal of Thermophysics*.

Important, comprehensive review articles are published in *Analytical Chemistry* at two yearly intervals.

1.B.2   Major Textbooks Devoted to Thermal Analysis and Calorimetry (in English)

1.B.2.1   Textbooks from 1950 to 1985

These early texts contain extremely valuable reference material, but regrettably are mostly out of print. Many scientific libraries will have copies, however.

C. Duval, *Inorganic Thermogravimetric Analysis*, Elsevier, Amsterdam, 1953 (2nd edn., 1963).

W. E. Garner (ed.), *Chemistry of the Solid State*, Butterworths, London, 1955.

N. Grassie, *Chemistry of High Polymer Degradation Processes*, Butterworths, London, 1956.

F. D. Rossini (ed.), *Experimental Thermochemistry I*, Wiley-Interscience, New York, 1956.

R. C. Mackenzie (ed.), *The Differential Thermal Analysis of Clays*, Mineralogical Society, London, 1957.

W. M. Smit (ed.), *Purity Control by Thermal Analysis*, Elsevier, Amsterdam, 1957.

W. J. Smothers and Y. Chiang, *Differential Thermal Analysis, Theory and Practice*, Chemical Publishing, New York, 1958.

C. G. Hyde and M. W. Jones, *Gas Calorimetry*, Ernest Benn, London, 1960.

G. N. Lewis and M. Randall, *Thermodynamics*, revised K. Pitzer, L. Brewer, McGraw-Hill, New York, 2nd edn., 1961.

H. A. Skinner (ed.), *Experimental Thermochemistry II*, Wiley, New York, 1962.

B. Ke (ed.), *Thermal Analysis of High Polymers*, Interscience, New York, 1964.

S. L. Madorsky, *Thermal Degradation of Organic Polymers*, Interscience, New York, 1964.

W. W. Wendlandt, *Thermal Methods of Analysis*, Interscience, New York, 1964.

P. D. Garn, *Thermoanalytical Methods of Investigation*, Academic Press, New York, 1965.

B. Pugh, *Fuel Calorimetry*, Butterworth, London, 1966.

R. F. Schwenker Jr. (ed.), *Thermoanalysis of Fiber and Fiber-forming Polymers*, Interscience, New York, 1966.

P. E. Slade and L. T. Jenkins (ed.) *Techniques and Methods of Polymer Investigation: Vol.* 1: *Thermal Analysis*, Arnold, London, 1966.

W. J. Smothers and Y. Chiang, *Handbook of Differential Thermal Analysis*, Chemical Publishing, New York, 1966.

W. Lodding, *Gas Effluent Analysis*, Marcel Dekker, New York, 1967.

J. P. McCullough and D. W. Scott (ed.), *Experimental Thermodynamics, Vol.*1: *Calorimetry of Non-reacting Systems*, Butterworths, London, 1968.

R. S. Porter and J. F. Johnson (ed.), *Analytical Calorimetry*, Plenum, New York, 1968.

H. J. V. Tyrell and A. E. Beezer, *Thermometric Titrimetry*, Chapman & Hall, London, 1968.

L. S. Bark and S. M. Bark, *Thermometric Titrimetry*, Pergamon Press, London, 1969.

C. J. Keattch, *An Introduction to Thermogravimetry*, Heyden, London, 1969. (2nd edn, with D. Dollimore, 1975).

V. S. Ramachandran, *Applications of Differential Thermal Analysis in Cement Chemistry*, Chemical Publishing, New York, 1969.

D. R. Stull, E. F. Westrum and G. C. Sinke, *The Chemical Thermodynamics of Organic Compounds*, Wiley, New York, 1969.

R. C. Mackenzie (ed.), *Differential Thermal Analysis*, Academic Press, London, 1970, Vol. 1, 1972, Vol. 2.

G. Liptay, *Atlas of Thermoanalytical Curves*, Heyden, London, 1971 to 1976, Vol. 1 to Vol. 5.

T. Daniels, *Thermal Analysis*, Kogan Page, London, 1973.

A. Blazek, *Thermal Analysis*, Van Nostrand, London, 1974.

H. Kambe and P. D. Garn (ed.), *Thermal Analysis: Comparative Studies of Materials*, Wiley, New York, 1974.

W. Smykatz-Kloss, *Differential Thermal Analysis: Applications and Results in Mineralogy*, Springer Verlag, Berlin, 1974.

J. Barthel, *Thermometric Titrations*, Wiley, New York, 1975.

B. LeNeindre and B. Vodar (ed.), *Experimental Thermodynamics, Vol. 2. Calorimetry of Non-Reacting Fluids*, Butterworths, London, 1975.

J. L. McNaughton and C. T. Mortimer, *Differential Scanning Calorimetry*, Perkin-Elmer, Reprinted from IRS Physical Chemistry series 2, Butterworth, London, 1975.

C. Duval, *Thermal Methods in Analytical Chemistry*, Elsevier, Amsterdam, 1976.

D. N. Todor, *Thermal Analysis of Minerals*, Abacus Press, Tunbridge Wells, 1976.

M. I. Pope and M. D. Judd, *Differential Thermal Analysis: A Guide to the Technique and its Applications*, Heyden, London, 1977.

S. Sunner and M. Mansson (ed.), *Experimental Chemical Thermodynamics, Vol. 1 Combustion Calorimetry*, Pergamon, Oxford, 1979.

M. E. Brown, D. Dollimore and A. K. Galwey, in *Comprehensive Chemical Kinetics*, ed. C. H. Bamford and C. H. Tipper *Vol. 22: Reactions in the Solid State*, Elsevier, Amsterdam, 1980.

A. E. Beezer (ed.), *Biological Microcalorimetry*, Academic Press, London, 1980.

E. A. Turi (ed.) *Thermal Characterization of Polymeric Materials*, Academic Press, New York, 1981.

C. Svehla (ed.), in *Wilson and Wilson's Comprehensive Analytical Chemistry, Vol. XII: Thermal Analysis, Parts A–D*, Elsevier, Amsterdam, 1981–1984.

I. M. Kolthoff, P. J. Elving and C. B. Murphy (ed.), *Treatise on Analytical Chemistry, Part I: Theory and Practice*, 2nd edn., *Vol. 12, Section J, Thermal Methods*, Wiley, New York, 1983.

K. J. Vorhees (ed.), *Analytical Pyrolysis*, Butterworth, London, 1984.

K. D. Maglic, A. Cezairliyan and V. E. Peletzky (ed.), *Compendium of Thermophysical Property Measurement Methods: Vol. 1: Survey of Measurement Techniques*, 1984, *Vol. 2: Recommended Measurement Techniques and Practices*, 1992, Plenum, New York.

C. M. Earnest, *Thermal Analysis of Clays, Minerals and Coal*, Perkin-Elmer, Connecticut, 1984.

W. Hemminger and G. Höhne, *Calorimetry–Fundamentals and Practice*, Verlag-Chemie, Weinheim, 1984.

1.B.2.2   Textbooks from 1985 to present

M. J. Aitkin, *Thermoluminescence Dating*, Academic Press, London, 1985.

K. Grime (ed.), *Analytical Solution Calorimetry*, Wiley, New York, 1985.

W. W. Wendlandt, *Thermal Analysis*, Wiley, New York, 3rd edn., 1986.

K. K. Kuo, *Principles of Combustion*, Wiley-Interscience, New York, 1986.

J. D. Pedley, R. D. Naylor and S. P. Kirby, *Thermochemical Data of Organic Compounds*, Chapman & Hall, London, 1986.

G. Widman and R. Riesen, *Thermal Analysis: Terms, Methods, Applications*, Huthig, Heidelberg, 1987.

J. W. Dodd and K. H. Tonge, *Thermal Methods*, (ACOL Series), Wiley, Chichester, 1987.

C. M. Earnest, *Compositional Analysis by Thermogravimetry*, ASTM, Philadelphia, 1988.

M. E. Brown, *Introduction to Thermal Analysis*, Chapman & Hall, London, 1988.

A. Cezairlyan (ed.), *CINDAS Data on Material Properties, Vol. 1-2. Specific Heat of Solids*, Hemisphere, New York, 1988.

P. Laye and M. Singh, *Hazard Evaluation by Thermal Analysis*, Health and Safety Executive, London, 1988.

J. A. McLean and G. Tobin, *Animal and Human Calorimetry*, Cambridge University Press, Cambridge, 1988.

J. L. Ford and P. Timmins, *Pharmaceutical Thermal Analysis: Techniques and Applications*, Ellis Horwood, Chichester, 1989.

V. R. Harwalker and C-Y. Ma (ed.), *Thermal Analysis of Foods*, Thomson Science, London, 1990.

B. Wunderlich, *Thermal Analysis*, Academic Press, New York, 1990.

E. Kaisersberger and H. Mohler, *DSC on Polymeric Materials*, Netzsch Annual for Science and Industry, Wurzburg, Selb, 1991, Vol. 1.

G. Lawson, *Chemical Analysis of Polymers*, RAPRA Review No. 47, RAPRA, Shawbury, 1991.

J. O. Hill, *For Better Thermal Analysis and Calorimetry*, ICTAC, New York, 3rd edn., 1991.

W. Smykatz-Kloss and S. St. J. Warne, *Thermal Analysis in the Geosciences*, Springer, Berlin, 1991.

G. van der Plaats, *The Practice of Thermal Analysis*, Mettler, 1991.

E. L. Charsley and S. B. Warrington (ed.), *Thermal Analysis: Techniques and Applications*, Royal Society of Chemistry, Cambridge, 1992.

J. D. Winefordner, D. Dollimore, J. Dunn and I. M. Kolthoff (ed.), *Treatise on Analytical Chemistry, Vol. 13. Part 1: Thermal Methods*, Wiley Interscience, New York, 2nd edn., 1993.

E. Kaisersberger, S. Knappe and H. Mohler, *TA for Polymer Engineering*, Netzsch Annual, Wurzburg, Selb, 1993, Vol. 2.

O. Kubaschewski, C. B. Alcock and P. J. Spencer, *Materials Thermochemistry*, Pergamon, Oxford, 6th edn., 1993.

V. A. Berstein and V. M. Egorov, *Differential Scanning Calorimetry of Polymers*, Ellis Horwood, Chichester, 1994.

T. Hatakeyama and F. X. Quinn, *Thermal Analysis: Fundamentals and Applications to Polymer Systems*, Wiley, Chichester, 1994, (2nd edn., 1999).

E. Kaisersberger, S. Knappe, H.Mohler and S. Rahner, *TA for Polymer Engineering*, Netzsch Annual, Wurzburg, Selb, 1994, Vol. 3.

J. M. Klotz and R. M. Posenberg, *Chemical Thermodynamics*, Wiley, New York, 1994.

R. F. Speyer, *Thermal Analysis of Materials*, Marcel Dekker, New York, 1994.

K. N. Marsh and P. A. G. O'Hare (ed.), *Solution Calorimetry, Vol. 4: Experimental Thermodynamics*, IUPAC, New York, 1994.

V. B. F. Mathot (ed.), *Calorimetry and Thermal Analysis of Polymers*, Carl Hanser, Munchen, 1994.

H. Brittain (ed.), *Physical Characterization of Pharmaceutical Solids*, Marcel Dekker, New York, 1995.

D. Q. M. Craig, *Dielectric Analysis of Pharmaceutical Systems*, Taylor & Francis, London, 1995.

P. J. Haines, *Thermal Methods of Analysis: Principles, Applications and Problems*, Blackie, Glasgow, 1995.

H. Kopsch, *Thermal Methods in Petroleum Analysis*, Wiley, Chichester, 1995.

F. Paulik, *Special Trends in Thermal Analysis*, Wiley, Chichester, 1995.

V. V. Boldyrev (ed.), *Reactivity of Solids*, Blackwell, Oxford, 1996.

G. Hohne, W.Hemminger and H-J. Flammersheim, *Differential Scanning Calorimetry – An Introduction for Practitioners*, Springer, Berlin, 1996.

J. Swarbrick and J. C. Boylan (ed.), *Encyclopaedia of Pharmaceutical Technology*, Vol. 15, *Thermal Analysis of Drugs and Drug Products*, Marcel Dekker, New York, 1996.

E. A. Turi (ed.), *Thermal Characterization of Polymeric Materials*, Academic Press, San Diego, 2nd edn., 1997, 2 vols.

J. P. Runt and J. J. Fitzgerald (ed.), *Dielectric Spectroscopy of Polymeric Materials: Fundamentals and Applications*, ACS, New York, 1997.

M. P. Sepe, *Thermal Analysis of Polymers*, RAPRA Review Report #95, RAPRA, Shawbury, 1997.

R. Brown (ed.), *Handbook of Polymer Testing: Physical Methods*, Marcel Dekker, New York, 1998.

M. E. Brown (ed.), *Handbook of Thermal Analysis and Calorimetry, Vol. 1: Principles and Practice*, Elsevier, Amsterdam, 1998.

J. E. Ladbury and B. Z. Chowdhry (ed.), *Biocalorimetry: Applications of Calorimetry in Biological Sciences*, Wiley, Chichester, 1998.

M. P. Sepe, *Dynamic Mechanical Analysis for Plastics Engineering*, Plastics Design Library/ William Andrew, New York, 1998.

T. Hatakeyama and L. Zhenhai, *Handbook of Thermal Analysis*, Wiley, Chichester, 1998.

J. L. Ford, P. Timmins and G. Buckton, *Pharmaceutical Thermal Analysis*, Taylor and Francis, London, 2nd edn., 1999.

R. B. Kemp (ed.), *Handbook of Thermal Analysis and Calorimetry, Vol. 4: From Molecules to Man*, Elsevier, Amsterdam, 1999.

T. M. Letcher (ed.), *Chemical Thermodynamics*, Blackwell Science, Oxford, 1999.

K. P. Menard, *Dynamic Mechanical Analysis: A Practical Introduction*, CRC Press, New York, 1999.

H. Utschick, *Methods of Thermal Analysis*, Ecomed, Perkin-Elmer, Landsberg, 1999.

Yi-Kan Cheng (ed.), *Electrothermal Analysis of VLSI Systems*, Kluwer, Amsterdam, 2000.

S. E. Harding and B. Chowdhry (ed.), *Protein–Ligand Interactions: Hydrodynamics and Calorimetry*, Oxford University Press, Oxford, 2000.

F. Kreith (ed.), *CRC Handbook of Thermal Engineering*, CRC, London, 2000.

R. Wigmans, *Calorimetry*, Oxford University Press, Oxford, 2000.

1.B.3 Videos, CD-Roms and other information

University of York, Thermal Analysis Videos, York Electronics Centre, York, UK (01904 432333), 1994.

Chemistry Video Consortium, Basic Laboratory Chemistry, Educational Media Film and Video, Harrow, UK (0208868 1908), 1997.

1.B.4 Other Literature and Learning Resources

Conference Proceedings. The proceedings of national and international conferences on thermal analysis and calorimetry are often published, either as independent books, or as Special Issues of the major Journals noted above.

International Congresses for Thermal Analysis and Calorimetry (ICTAC)

1st ICTA: Aberdeen, 1965, J. P. Redfern (ed.), Macmillan, London, 1965.

2nd ICTA: Worcester, 1969, R. F. Schwenker and P. D. Gard (ed.), 2 vols., Academic Press, New York, 1969.

3rd ICTA: Davos, 1972, H. G. Wiedemann (ed.), 3 vols., Birkhauser, Basel, 1972.

4th ICTA: Budapest, 1975, I. Buzas (ed.), 3 vols., Heyden, London, 1975.

5th ICTA: Kyoto, 1977, H. Chihara (ed.), Heyden, London, 1977.

6th ICTA: Bayreuth,1980, H. G. Wiedemann and W. Hemminger (ed.), Birkhauser, Basel, 1980.

7th ICTA, Kingston, Canada, 1982, B. Miller (ed.), 2 vols., Wiley, New York, 1982.

8th ICTA, Bratislava, 1985, A. Blazek (ed.), *Thermochim. Acta*, 1985, **92–93**.

9th ICTA, Jerusalem, 1988, S. Yariv (ed.), *Thermochim. Acta*, 1988, **133–135** and **148**.

10th ICTA, Hatfield, 1992, D. J. Morgan (ed.), *J. Thermal Anal.*, 1993, **40**, (3 vol.).

11th ICTAC, Philadelphia, M. Y. Keating (ed.), *J. Thermal Anal.*, 1997, **49**, (3 vol.).

12th ICTAC, Copenhagen, 2000, O. T. Sørensen and P. J. Møller (ed.), *J. Therm. Anal. Cal.*, 2001, **64**, (3 vol.).

European Symposia on Thermal Analysis and Calorimetry (ESTAC)
1st ESTA, Salford, 1976, D. Dollimore (ed.), Heyden, London, 1976.

2nd ESTA, Aberdeen, 1981, D. Dollimore (ed.), Heyden, London, 1981.

3rd ESTAC, Interlaken, 1985, E. Marti and H. R. Ostwald (ed.), *Thermochim. Acta*, 1985, **85**.

4th ESTAC, Jena, 1988, D. Schultze (ed.), *J. Thermal Anal.*, 1988, **33**.

5th ESTAC, Nice, 1992, R. Castanat and E. Karmazsin (ed.), *J. Thermal Anal.*, 1992, **38**.

6th ESTAC, Grado, 1994, A. Cesaro and G. Della Gatta (ed.), *Thermochim. Acta*, 1995, **269–70**.

7th ESTAC, Balatonfured, Hungary, 1998, J. Kristof and C. Novak (ed.), *J. Therm. Anal. Cal.*, 1998.

International Union of Pure and Applied Chemistry (IUPAC) : Conferences on Chemical Thermodynamics (since 1990)
11th Conference, Como, Italy, 1991, *Pure Appl. Chem.*, 1991, **63**.

12th Conference, Snowbird, Utah, 1993, *Pure Appl. Chem.*, 1993, **65**.

13th Conference, Clermont-Ferrand, France, 1994, *Pure Appl. Chem.*, 1995, **67**.

APPENDIX 2   ICTAC AND ITS AFFILIATED
SOCIETIES IN EUROPE AND THE USA

ICTAC

Membership Secretary, Dr H. G. McAdie, 104 Golfdale Road, Toronto, Ontario, Canada, M4N 2B7
E-mail: hgmcadie@aol.com
Web: http://www.ictac.org

Czech Working Group on Thermal Analysis

Professor J. Sestak, Academy of Sciences, Institute of Physics, 10, Cukrovamickastr, 162 00 Prague 6, Czech Republic
E-mail: sestak@fzu.cz

France: AFCAT

Professor E. Karmazin, 17 Montee Saint Sebastien, 69001 Lyon, France
E-mail: karmazin@cpe.fr

Germany: GEFTA

Dr M Feist, Humboldt Universitat zu Berlin, Institut fur Chemie, Hessischestrasse 1-2, D-10115 Berlin, Germany
E-mail: feistm@chemie.hu.berlin.de

Greece: HSTA

Professor G. Parissakis, National Technical University of Athens, Department of Chemical Engineering, 9 Heroon Polytechniou Street, GR-15773 Athens, Greece

Hungary: HUNGTAG

Professor J. Simon, Lexica Ltd, Amerikai ut 52, 1145 Budapest, Hungary
E-mail: lexical@mail.datanet.hu

Italy: AICAT

Professor A. Buri, Dipartimento di Ingereria del Materiali e della Produzioni, Universita degli Studi di Napoli, Federico II, DIMP, Piazzale Tecchio, 81025 Napoli NA, Italy

Netherlands: TAWN

Dr V. B. Mathot, DSM Research PAC-MC, PO Box 18, 6160-MD, Geleen, Netherlands
E-mail: v.b.f.mathot@research.dsm.ne

North American Thermal Analysis Society: NATAS
Dr D.J. Burlett, Goodyear Tire and Rubber Co., 142 Goodyear Blvd,
Akron, OH 44305, USA
E-mail: djburlett@goodyear.com

Poland: TA Section of Polish Mineralogical Society
Professor Dr L. Stoch, University of Mining, Faculty of Materials
Science and Ceramics, ul. Mickiewicza 30, 30-059 Krakow, Poland

Russia: National Committee of Russia on Thermal Analysis and Calorimetry
Professor A. D. Izotov, Kurnakov Institute of General and Inorganic
Chemistry, Leninskii Prospect 31, 117907 Moscow, Russia

Scandinavia: Nordic Society for Thermal Analysis and Calorimetry: NOS-TAC
Dr G. Mogensen, Research & Development, Haldor Topsoe A/S,
Nymolievej 55, DK-2800 Lyngby, Denmark
E-mail: gum@htas.dk

Spain: Grupo De Calorimetria y Analisis Termica
Professor R. Nomen, Institut Quimic de Sarria, Via Augusta 390, 08017,
Barcelona, Spain
E-mail: rnome@iqs.url.es

Switzerland: SGTK
Dr E. Marti, Novartis Serv.AG, Klybeck K-127.556, CH-4002, Basel,
Switzerland
E-mail: Erwin.marti@sn.novartis.com

United Kingdom: TMG
Dr R. J. Willson,
GlaxoSmithKline, New Frontiers Science Park (South), Harlow, Essex,
CM19 5AW, UK
E-mail: Richard_Willson@sbphrd.com

APPENDIX 3   AMERICAN AND OTHER STANDARD TEST METHODS

Some of the national and international standards for thermal analysis
and calorimetry are given below, together with contact addresses for the
organisations.

3.1 American Society for Testing and Materials (ASTM) Standards

ASTM, 100 Barr Harbor Drive, West Consohocken, Pennsylvania 19428-2959, USA
Web: http://www.astm.org. The Standards may be downloaded from this site on payment of a fee.

The ASTM has a Committee (E-37) which deals particularly with Thermal Measurements and another (D-20) which concentrates on Plastics

C 351 (1992):	Test method for mean specific heat of thermal insulation
D 240 (1992):	Test method for heat of combustion of liquid hydrocarbon fuels by bomb calorimetry
D 696 (1998):	Standard Test Method for Coefficient of Linear Expansion of Plastics between $-30\,^{\circ}$C and $30\,^{\circ}$C with a Vitreous Silica Dilatometer
D 1519 (2000):	Test method for rubber chemicals – melting range
D 1826 (1994):	Test method for calorific (heating) value of gases in natural gas range by continuous recording calorimeter
D 1989 (1995):	Test method for gross calorific value of coal and coke by microprocessor controlled isoperibol calorimeters
D 2015 (1995):	Test method for gross calorific value of coal and coke by the adiabatic bomb calorimeter
D 2382 (1988):	Test method for heat of combustion of hydrocarbon fuels by bomb calorimeter (high-precision method)
D 2766 (1995):	Test method for specific heats of liquids and solids
D 3286 (1991):	Test method for gross calorific value of coal and coke by the isoperibol bomb calorimeter
D 3350 (1999):	Polyethylene Pipes and Fitting Materials
D 3386 (1994):	Test method for coefficient of linear thermal expansion of electrical insulating materials
D 3417 (1999):	Test method for heats of fusion and crystallization of polymers by thermal analysis
D 3418 (1999):	Test method for transition temperatures of polymers by thermal analysis
D 3850 (1994):	Test method for rapid thermal degradation of solid electrical insulating materials by thermogravimetric method (TGA)
D 3895 (1998):	Test method for oxidative-induction time of polyolefins by differential scanning calorimetry

D 3947 (1992):   Test method for specific heat of aircraft turbine lubricants by thermal analysis

D 4000 (1995):   Identification of Plastic Materials

D 4065 (1995):   Determining and reporting dynamic mechanical properties of plastics

D 4092 (1996):   Terminology relating to dynamic mechanical measurements on plastics

D 4419 (1995):   Test method for measurement of transition temperatures of petroleum waxes by differential scanning calorimetry (DSC)

D 4535 (1985):   Test method for thermal expansion of rock using a dilatometer

D 4565 (1999):   Physical and environmental performance properties of insulations and jackets for telecommunication wire and cable

D 4591 (1997):   Temperatures and heats of transition of fluoropolymers by DSC

D 4611 (1986):   Test method for specific heat of rock and soil

D 4809 (1995):   Test method for heats of combustion of liquid hydrocarbon fuels by bomb calorimetry (precision method)

D 4816 (1994):   Test method for specific heat of aircraft turbine fuels by thermal analysis

D 5028 (1996):   Test method for curing properties of pultrusion resins by thermal analysis

D 5468 (1995):   Standard test method for gross calorific and ash value of waste materials

D 5483 (1995):   Test method for oxidation induction time of lubricating greases by pressure differential scanning calorimetry

D 5865 (1995):   Test method for gross calorific value of coal and coke

D 5885 (1997):   Test method for oxidation induction time of polyolefin geosynthetics by high-pressure differential scanning calorimetry

D 6186 (1998):   Oxidation Induction Time of Lubrication Oils by Pressure DSC

D 6370 (1999):   Standard Test Method for Rubber-Compositional Analysis by thermogravimetry

D 6375 (1999):   Standard Test Method for Evaporation Loss of Lubricating Oils by TGA, Noack Method

D 6382 (1999):	Standard Practice for Dynamic Mechanical Analysis of Thermal Expansion of Rock using a Dilatometer
D 6546 (2000):	Standard Test Method for Determining the Compatibility of Elastomer Seals
D 6558 (2000):	Standard Test Method for Determination of TGA CO_2 Reactivity of Baked Carbon Anodes and Cathode Blocks
D 6559 (2000):	Standard Test Method for Determination of TGA Air Reactivity of Baked Carbon Anodes and Cathode Blocks
E 228 (1995):	Linear thermal expansion of solid materials with a vitreous silica dilatometer
E 289 (1995):	Linear thermal expansion of rigid solids with interferometry
E 422 (1994):	Test method for measuring heat flux using a water-cooled calorimeter
E 472 (1991):	Reporting thermoanalytical data
E 473 (1999):	Terminology relating to thermal analysis
E 476 (1993):	Standard test method for thermal instability of confined condensed phase systems (confinement test)
E 487 (1999):	Test method for constant-temperature stability of chemical materials
E 537 (1998):	Test method for assessing the thermal stability of chemicals by methods of differential thermal analysis
E 698 (1999):	Test method for Arrhenius kinetic constants for thermally unstable materials
E 711 (1992):	Test method for gross calorific value of refuse-derived fuel by the bomb calorimeter
E 793 (1995):	Test method for heats of fusion and crystallization by differential scanning calorimetry
E 794 (1998):	Test method for melting and crystallization temperatures by thermal analysis
E 831 (2000):	Test method for linear thermal expansion of solid materials by thermomechanical analysis
E 928 (1996):	Test method for mol percent impurity by differential scanning calorimetry
E 967 (1999):	Standard practice for temperature calibration of differential scanning calorimeters and differential thermal analyzers

E 968 (1999):	Standard practice for heat flow calibration of differential scanning calorimeters
E 1131 (1998):	Test method for compositional analysis by thermogravimetry
E 1142 (1997):	Terminology relating to thermophysical properties
E 1231 (1996):	Calculation of hazard potential figures-of-merit for thermally unstable materials
E 1269 (1999):	Test method for determining specific heat capacity by differential scanning calorimetry
E 1354 (1994):	Test method for heat and visible smoke release rates for materials and products using an oxygen consumption calorimeter
E 1356 (1998):	Test method for glass transition temperatures by differential scanning calorimeter or differential thermal analysis
E 1363 (1997):	Temperature calibration of thermomechanical analyzers
E 1545 (2000):	Standard test method for glass transition temperatures by thermomechanical analysis
E 1559 (1993):	Contamination outgassing characteristics of spacecraft materials
E 1582 (1993):	Standard practice for calibration of temperature scale for thermogravimetry
E 1640 (1999):	Test method for assignment of the glass transition temperature by dynamic mechanical analysis
E 1641 (1999):	Test method for decomposition kinetics by thermogravimetry
E 1782 (1998):	Vapour pressure by DSC/DTA
E 1824 (1996):	Standard Test Method for Assignment of a Glass Transition Temperature using Thermomechanical Analysis under Tension
E 1858 (1997):	Oxidative Induction Time by DSC
E 1867 (1997):	Standard Test Method for Temperature Calibration of Dynamic Mechanical Analysers
E 1868 (1997):	Loss on Drying by TGA
E 1877 (2000):	Thermal Endurance from TGA Decomposition Data
E 1952 (1998):	Thermal Diffusivity/Conductivity by MTDSC
E 1953 (1998):	Description of Thermal Analysis Apparatus
E 1970 (1998):	Statistical Treatment of Thermal Analysis Data
E 1981 (1998)	Guide for assessing the thermal stability of materials by the method of Accelerating Rate Calorimetry

E 2008 (1999):   Volatility Rate by TGA

E 2009 (1999):   Oxidation Onset Temperature by DSC

E 2038 (1999):   Standard Test Method for Temperature Calibration of Dielectric Analysers

E 2039 (2000):   Standard Practice for Determining and Reporting Dynamic Dielectric Properties

E 2040 (2000):   Mass Calibration of TGA

E 2041 (1999):   Borchardt & Daniels Kinetics by DSC

E 2046 (2000):   Reaction Induction Time by DSC

E 2069 (2000):   DSC Calibration by Cooling

E 2070 (2000):   Isothermal Kinetics by DSC

E 2071 (2000):   Heat of Vaporization

E 2092 (2000):   Standard Test Method for Distortion Temperature in Three-Point Bending by Thermomechanical Analysis

E 2113 (2000):   TMA Length Change Calibration

3.2 British Standards

BS 2000-12 (1993):   Methods of test for petroleum and its products. Determination of specific energy

BS 3804-1 (1964):   Methods for the determination of the calorific values of fuel gases. Non-recording methods (obsolescent)

BS 4550-3 (1978):   Methods of testing cement. Physical tests. Test for heat of hydration

BS 4791 (1985):   Specification for calorimeter bombs

BS 7420 (1991):   Guide for determination of calorific values of solid, liquid and gaseous fuels (including definitions)

BS EN ISO 11357 – Part I: Plastics – Differential scanning calorimetry – General principles

BS EN ISO 11357-2 (1999):   Plastics – Differential scanning calorimetry – Determination of glass transition temperature

BS EN ISO 11357-3 (1999):   Plastics – Differential scanning calorimetry – Determination of temperature and enthalpy of melting and crystallisation

BS EN ISO 11357-5 (2000):   Plastics – Differential scanning calorimetry – Determination of characteristic reaction-curve temperatures and times, enthalpy

	of reaction and degrees of conversion
ISO/FDIS 11357 – Part 4:	Plastics – Differential scanning calorimetry – Determination of specific heat capacity
ISO/DIS 11357 – Part 6:	Plastics – Differential scanning calorimetry – Determination of oxidation induction time
ISO/DIS 11357 – Part 7:	Plastics – Differential scanning calorimetry – Determination of crystallization kinetics
ISO/CD 11357 – Part 8:	Plastics – Differential scanning calorimetry – Determination of amount of water absorbed by polymers
ISO/CD 11358 – Part I:	Plastics – Thermogravimetry of Polymers – General principles
ISO/CD 11358 – Part 2:	Plastics – Thermogravimetry of Polymers – Determination of kinetic parameters
BS ISO 11359 – Part (1999):	Plastics – Thermomechanical analysis (TMA) – General principles
BS ISO 11359 – Part 2 (1999):	Plastics – Thermomechanical analysis (TMA) – Determination of linear thermal expansion coefficient and glass transition temperature
ISO/FDIS 11359 – Part 3:	Plastics – Thermomechanical analysis (TMA) – Determination of softening temperature
BS EN ISO 6721 – Part I:	Plastics – Determination of dynamic mechanical properties – General principles
NWI/ISO 6721 – Part 11:	Plastics – Dynamic mechanical analysis – Determination of glass transition temperature
NWI/ISO 6721 – Part 12:	Plastics – Dynamic mechanical analysis – Calibration

Further Information from Graham Sims, National Physical
Laboratory, Teddington, Middlesex, UK, TW1 1 OLW
Tel: 020 8943 6564; Fax: 020 8614 0433.

APPENDIX 4   MANUFACTURERS, CONSULTANTS AND OTHER SUPPLIERS

Many of the firms listed below provide comprehensive brochures, Application Notes and Internet information on their products. Many also produce informative material on the principles of the techniques employed.

BÄHR-Thermoanalyse GmbH
PO Box 1105, D-32603 Hüllhorst, Germany
Fax: 05744-1006
Web: http://www.baehr-thermo.de
E-mail: info@baehr-thermo.de
(Simultaneous thermal analysers, *etc.*)

Bohlin Instruments
5 Love Lane, Cirencester, Gloucestershire GL7 1YG, UK
Fax: 01285 644314
E-mail: lw@bohlin.co.uk
(Rheometers)

Brookhaven Instruments Ltd
Chapel House, Stock Wood, Worcestershire B96 6ST, UK
Fax: 01386 792720
Web: http://www.brookhaven.co.uk/thanda.html
E-mail: info@brookhaven.co.uk
(SETARAM Instruments, *etc.*)

Cahn Instruments
5225, Verona Road, Bldg 1, Madison, WI 53711, USA
Fax: (608) 273-6827
Web: http://www.cahn.com
E-mail: info@cahn.com
(TG systems)

Calorimetry Sciences Corp.
155 West 2050 North, Spanish Fork, Utah 84660, USA
Fax: (801) 794 2700
Web: http://www.calorimetrysciences.com/
(Calorimeters)

CI Electronics Ltd
Brunel Road, Churchfields, Salisbury, Wiltshire SP2 7PX, UK
Fax: 01722 323222
E-mail: admin@cielec.com
(Microbalances)

Columbia Scientific Industries Corp.
11950 Jollyville Road, Austin, Texas 78759, USA
Fax: (512) 258 5004
(Reaction calorimeters)

European Spectrometry Systems Ltd (ESS)
GeneSys House, Denton Drive, Northwich, Cheshire CW9 7LU, UK
Fax: 01606 330937
E-mail: service@essco.u-net.com
(Simultaneous Thermal Analysis (STA) and Mass Spectrometry Systems)

The Fine Work Co. Ltd
West Mead House, 123, West Mead Road, Sutton SM1 4JH, UK
Fax: 0208 770 3911
Web: http://www.fwco.demon.co.uk
E-mail: Fine@fwco.demon.co.uk
(Scientific instrument assemblies)

Gearing Scientific
1 Ashwell Street, Ashwell, Hertfordshire SG7 5QF, UK
Fax: 01462 742565
(Dynamic mechanical analysers, contract thermal analysis)

Hazard Evaluation Laboratory Ltd (HEL)
50 Moxon Street, Barnet, Hertfordshire EN5 5TS, UK
Fax: 0208 441 6754
Web: http://www.helgroup.co.uk/
E-mail: info@helgroup.co.uk
(Calorimetric testing)

Heath Scientific Company Ltd
1 North House, Bond Ave, Bletchley MK1 1SW, UK
Fax: 01908 645209
Web: http://www.science.org.uk/tht-1.htm
E-mail: science@easynet.co.uk
(Thermal hazard testing, calorimeters, spectrometers)

Hiden Analytical Ltd
420 Europa Boulevard, Warrington WA5 5UN, UK
Fax: 01925 416518
Web: http://www.hidenanalytical.com/start.html
E-mail: sales@hiden.demon.co.uk
(Thermogravimetric analysers, *etc.*)

IKA-Labortechnik
Jankel & Kunkel-Str., 10, d-79219, Staufen, Germany
Fax: 07633 831-98
E-mail: sales@IKA.de
(Calorimeters, *etc.*)

Inframetrics Infrared Systems Ltd
Riverside Studio, Buxton Road, Bakewell DE45 1GS, UK
Fax: 01629 814400
E-mail: infrauk@globalnet.co.uk
(Infrared imaging systems)

Infra Scientific Ltd
Unit 11, Station Court, Park Mill Way, Clayton West, West Yorkshire
HD8 9XJ, UK
Fax: 01484 864882
Web: http://www.infra.uk.com
E-mail: info@infra.uk.com

Instron Ltd
Coronation Road, High Wycombe, Buckinghamshire HP12 3SY, UK
Fax: 01494 6456123
(Instruments for materials testing)

Instrument Specialists
2402 Spring Ridge Drive, Suite B, Spring Grove, IL 60081, USA
Fax: (815) 675-1552
Web: http://www.instrument-specialists.com
E-mail: isi@instrument-specialists.com
(Thermal analysis instrumentation)

Kinetica Inc.
9562 North Dixie Highway, Franklin OH 45005, USA
Web: http://www.thermochemistry.com/omdex.html
E-mail: kinetica@thermochemistry.com
(Calorimetry, hazard testing)

Linkam Scientific Instruments
8 Epsom Downs Metro Centre, Waterfield, Tadworth KT20 5HT, UK
Fax: 01737 363480
Web: http://www.linkam.co.uk
E-mail: info@linkam.co.uk
(Hot-stage microscopes and accessories)

Linseis GmbH
15 St. Peters Street, Duxford, Cambridge CB2 4RP, UK
Fax: 01223 837367
Web: http:www.linseis.com/thermal.htm
(Thermal analysis equipment and accessories)

Magna Projects and Instruments Ltd
17 Braemar Close, Mountsorrel LE12 7ES, UK
Fax: 0116 230 3606
E-mail: magna@recarter.demon.co.uk
(Viscometry)

Mathis Instruments
PO Box 69000, 8 Garland Court, Incutech Bldg, Suite 207, Fredericton,
New Brunswick, Canada, E3B 6C2
Fax: + 1 506 462 7210
Web: http://www.mathisinstruments.com
E-mail: info@MathisInstruments.com
(Thermal conductivity instruments: see also PETA)

Mettler Toledo Ltd
64, Boston Road, Beaumont Leys, Leicester LE4 1AW, UK
Fax: 0116 235 0888;
Web:http://www.mt.com/home/products/en/lab/thermal/analysis.asp
E-mail: enquire@mtuk.mt.com
(Thermal analysis instruments *etc.*)

MicroCal
22 Industrial Drive East, Northampton, Massachusetts 01060-2327, USA
Fax: (413) 586 0149
Web: http://www.microcalorimetry.com/
E-mail: info@microcalorimetry.com
(Calorimeters)

Netzsch-Mastermix Ltd
Thermal Analysis Section, Vigo Place, Aldridge, Walsall WS9 8UG, UK
Fax: 01922 453320
Web: http://www.ngb.netzsch.com
E-mail: ngb.nmx@netzsch.com
(Thermal analysis instruments *etc.*)

Novocontrol International
PO Box 63, Worcester WR2 6NJ, UK
Fax: 01905 642642
(Dielectric thermal analysers, dielectric spectrometers)

Paar Scientific Ltd
594 Kingston Road, Raynes Park, London SW20 8DN, UK
Fax: 0208543 8727
E-mail: paar@psl.anton-paar.co.uk
(Temperature controlled density and rheometric equipment)

PETA Solutions (Perkin-Elmer)
PO Box 188, Beaconsfield, Buckinghamshire HP9 2GB, UK
Fax: 01494 679280
Web: http://www.thermal-instruments.com
E-mail: gabbotpv@perkin-elmer.com
(Thermal analysis instruments *etc.*)

RAPRA Technology Ltd
Shawbury, Shrewsbury SY4 4NR, UK
Fax: 01939 251118
E-mail: info@rapra.net
(Consultancy and publications on rubbers and polymers)

Rheometric Scientific
One Possumtown Road, Piscataway, NJ 08854, USA
Fax: (732) 560 7451
Web: http://www.rheosci.com/

Rigaku Industrial Corp.
E-mail: info@rigaku.co.jp
(Thermal analysis equipment)

Rupprecht & Patashnick Co. Inc.
25 Corporate Circle, Albany, NY 12203, USA
Fax: (518) 452 0067
E-mail: info@rpco.com
(Microreactors, pulsed mass analysers)

Scientific and Medical (S&M)
Shirley House, 12, Gatley Road, Cheadle, Cheshire SK8 1PY, UK
Fax: 0161 428 7521
E-mail: scimed@dircon.co.uk
(Cahn balances and other thermal equipment)

Seiko Instruments Inc. (See also Carl Stuart)
Fax: (043) 211-8067
(Thermal analysis equipment)

SETARAM (see also Brookhaven)
7 rue de l'Oratoire, 69300, Caluire, France
Fax: (33) (0) 4 7828 6355
Web: http://www.setaram.fr

Shimadzu Europa (UK)
Mill Court, Featherstone Road, Wolverton Mill South, Milton Keynes
MK12 5RE, UK
Fax: 01908 552211
E-mail: sales@shimadzu.com
(Thermal analysis and other instrumentation)

Solomat Instrumentation
Glenbrook Industrial Park, Stamford, Ct 06906, USA
Fax: (203) 356-0125
(Calorimeters, TSC apparatus)

Spectra SensorTech Ltd
Cowley Way, Weston Road, Crewe CW1 6AG, UK
Fax: 01270 251939
(Mass spectrometric analysers)

Carl Stuart Ltd
Unit 20/21, Town Yard Business Park, Leek, Staffordshire ST13 8BF,
UK
Fax: 01538 398200
E-mail: carlstuart@fenetre.co.uk
(Thermal analysis equipment)

TA Instruments Ltd
Europe House, Bilton Centre, Cleeve Road, Leatherhead KT22 7UQ, UK
Fax: 01372 360135
Web: http://www.tainst.com
E-mail: info@taeurope.co.uk
(Thermal analysis, micro-thermal analysis and rheology products)

Thermold
652 Glenbrook Road, Stamford, CT 06906, USA
Fax: (203) 977 8237
Web: http://www.thermoldlp.com
E-mail: info@thermoldlp.com
(Calorimetry equipment)

Thermometric Ltd
10 Dalby Court, Gadbrook Business Park, Northwich, Cheshire
CW9 7TN, UK
Fax: 01606 48924
Web: http://www.thermometric.com
(Calorimetry equipment)

Triton Technology Ltd
Mansfield i-Centre, Oakham Business Park, Hamilton Way, Mansfield
NG18 5BR, UK
Fax: 01623 600626
Web: http://www.welcome.to/dynamic_mechanical_analysis
E-mail: enquiries@ttech.globalnet.co.uk
(Dynamic mechanical analysis instruments)

ULVAC-Sinku Riko
1-9-19 Hakusan, Midori-ku, Yokohama 226, Japan
Fax: (045) 933 9973
Web: www.ulvac-riko.co.jp
(Thermal analysis equipment)

APPENDIX 5   CONSULTANCIES AND OTHER GROUPS

Anasys Thermal Methods Consultancy
IPTME, Loughborough University, Loughborough LE11 3TU, UK
Fax: 01509 223332
Web: http://www.anasys.co.uk
E-mail: info@anasys.co.uk

Advanced Thermal AnalysiS (ATHAS)
Department of Chemistry, University of Tennessee, Knoxville, TN
37996-1600, USA
Fax: (615) 974 3454
Web: http://www.utk.edu/~athas/
E-mail: athas@utk.edu

B&K Publishing
PO Box 2868, West Lafayette, IN 47996, USA
Web: http://www.bkpublishing.com
E-mail: kkociba@bkpublishing.com
(Thermal analysis, calorimetry and rheology buyers guide)

Centre for Thermal Studies
University of Huddersfield, Queensgate, Huddersfield HD1 3DH, UK
Fax: 01484 473179
Web:http://www.hud.ac.uk/schools/applied-sciences/chem./cts.htm
E-mail: e.l.charsley@hud.ac.uk

National Physical Laboratory (NPL) Thermal Metrology Group
National Physical Laboratory, Teddington TW11 OLW, UK
Web: http://www.npl.co.uk/npl/science/thermal.html

OCAMAC (Oxford Centre for Advanced Materials And Composites)
Oxford University, Parks Road, Oxford OX1 3PH, UK
Fax: 01865 848790
Web: http://www.materials.ox.ac.uk/ocamac
E-mail: ocamac.director@materials.ox.ac.uk

Polymer Research Centre
University of Surrey, Guildford, Surrey GU2 5XH, UK
Fax: 01483 300803
Web:http://www.surrey.ac.uk/PRC/
E-mail: g.stevens@surrey.ac.uk

Thermal: Thermal Analysis Online
Contact: M. J. Rich, Composite Materials and Structure Center, Michigan State University, East Lansing, MI 48824-1326, USA
Fax: (517) 432 1634
Web: http://www.egr.msu.edu/mailman/listinfo/thermal
E-mail: rich@egr.msu.edu

Thermophysical Properties World
Eclectic Media Services, 1567 E. Algonquin Road, Des Plaines, IL 60016-6629, USA
Fax: 847 635 9170
Web: http://www.eclecticmedia.com/contact.htm
E-mail: cdoonan@eclecticmedia.com

Subject Index